THE POLITICS OF UNTOUCHABILITY

OWEN M. LYNCH

The Politics
of Untouchability

SOCIAL MOBILITY AND SOCIAL

CHANGE IN A CITY OF INDIA

COLUMBIA UNIVERSITY PRESS 1969

New York and London

Owen M. Lynch is associate professor
of Anthropology, State University
of New York at Binghamton.

For

JOHN, father   DAVID, brother   ESTHER, friend

FOREWORD

In his Acknowledgments, Owen Lynch reports what a deep personal experience of friendship, humanity, and the resilience of the human spirit his years in India and his year and a half with the Jatavs of Agra City provided. His book about their lives, their quest for new identities, and their rise despite their poverty and their earlier dependence to new citizenship within their city and nation is both a witness for them and controlled social science. It is also a most significant documentation of cultural and social transformation. It tells of political evolution, of the tides of growth and change, of the currents sweeping the new nations, and among them India, old, burdened, and various beyond the lot of most of them (though probably wiser thereby), moving into modernity.

The social science Owen Lynch represents in this book is anthropology. Cultural and social anthropology goes forward while still much linked to the natural history which gave it birth. Its chief method is immersion in a scene of field-work; its techniques are participant observation, open interviews, documentation from people of the reasons, the courses, and the personal involvements of their on-going lives and unfolding careers. This method was once put to recording tribes of preliterate peoples; it was later naturalized to villages and countryfolk, especially in India, by European and Indian anthropologists. It has only recently been further transferred to city wards and the new urban populations of the developing and developed worlds. Empirical, comparative, and direct, such social science gives rich detail of the present and the immediate while it attempts as well to explain what it finds in terms general enough to cover a spectrum of the societies of the world and of the changes of all human evolution, thus uniting behavior to the universal process. Thus the Camars of Agra City speak here not only for themselves but also, through Owen Lynch, for Indian society and civilization. Moreover, they also speak through him for world urbanization and political emergence, for universal stringencies of social stratification, groups and identities, and poverty and aspiration in society.

A good deal of this book, therefore, uses and discusses, and, I think one can fairly say, makes new contributions to wider theory. Cultural anthropology shares ground with sociology and political science: it overlaps also, less extensively, economics and history. Anthropological research done in literate countries of modern national form, seats of the high civilization of record keeping, of censuses and voting records, and the products of the printing presses, cannot rely only upon fieldwork. It needs more than the observations, interviews, and biographies gathered by sight and word of mouth between fieldworker and native, however rich the local contact and however faithful (even with tape recorders of today) the presentation of the fieldwork. It must also make judicious and economical use of the documentary evidence of local and national history and of the political or other data of local, national, and international movements. Owen Lynch, thus, has not hesitated to test out and sort among the views of those sociologists who have dealt with caste and with westernization. Two of them to whose theories he adds new sharpness and some extension are M. N. Srinivas, Indian master of sociology and anthropology who has codified the major dynamics of social change in India in terms reaching beyond her to other modernizing cultures, and Robert Merton, who has codified modern structural functionalism, reference group theory and role theory, keys to our understanding of social stratification in complex cultures everywhere.

Lynch's test of these men's contributions is the professional reward his colleagues will derive from reading him, matching well the personal reward others will find in the Camars' human story. It will provide cross-referencing to other societies and to the theoretical literature that will interest those who lack a firm background in Indian studies. It will thus help to bring the burgeoning, remarkably fertile new Indian sociology and anthropology, so greatly accelerated since Independence, into international and American attention. Those who have not read Wiener and Bailey and Dube in political science and Dumont, Beteille, Beals, and McKim Marriot in sociology, to name just a few Indian, European, and American scholars, do not realize what this rich fund of new experience can contribute to the growth of social science. Lynch's account of the Jatavs of Agra is couched broadly enough to be of considerable value to all students of mobility and of acculturation.

Quite specifically, the present work tells us a good deal about the nature of caste in general and the role of caste in social change in modern India in particular. M. N. Srinivas* remarks that "in traditional India, fission seems to have been the dominant process (in caste), whereas today . . . fusion has replaced fission." He is sure that fundamental change is occurring in the "Westernization" of India, in the emergence of caste blocs and in increasing political and social mobilization. These attest to the spread of egalitarian ideology depicted in Lynch's recording of the naturalization among former Untouchables in their new role of citizen. Srinivas feels that this change cannot be described as "a simple movement from a closed to an open system of social stratification." Owen Lynch has carried this perception of Srinivas' forward. The continuation of caste through new individual and familial mobility, through the creation of new political spokesmen, through the redefinitions of identity in old and new symbolisms, show us an acculturation of the old institutions of caste to the new ones of nationhood, which teach us that modernization proceeds as much by accommodating old and new together as by sweeping the old aside.

The book, then, is one of the fruits of a project, generously supported by the National Institute of Mental Health, carried out at Columbia University, in which we undertook to study initial industrialization, in an effort to identify persistences and transformations in the wake of world-wide modernizing pressures. The course of development may well run differently in, say, Japan and India from the way it ran in Britain or in Germany or in Russia.

* *Social Change in Modern India* (Berkeley, University of California Press, 1966) p. 115.

Owen Lynch reports on the leather-workers, moving from handi-
craft to factory production, not necessarily abandoning, but rather
expanding, their ancient ways and their ancient place in their
society. Others of the project, who began work in other parts of
India, will soon report in their works other persistences and
expansions, analogous to those of other countries, Japan, for
example, which may show that India like other civilizations can advance into modernity without ceasing to continue in her own
tradition. The special poignancy of the Untouchable case is that
as with American Negroes their ancient place was not a happy or
a voluntary one, and the confining attitudes of the larger society
that relegated Untouchables or Negroes to their places were not
of their own liking or choosing. Yet, nonetheless, in such social
change and new acculturation to the equalizing forces of modern
economic and political institutional structures as is reported here, a
transformation of poverty and inferiority into at least somewhat
greater range of earning and of occupation and into somewhat
stronger citizenship through the ballot is demonstrably and historically possible. David Mandelbaum, another American anthropologist deeply experienced in India, has written in a recent review
of a very different, but parallel book, *India's Ex-Untouchables,**
by Harold R. Isaacs, "The political changes in India (since Independence) have opened up new possibilities for the lowest and
poorest, opportunities that require changes in self-identity as well
as in identification by others."

  In a day when much is being said in anthropology and in urban
studies of the disadvantaged, the once-segregated poor, the hope-
lessness of the people caught in the "culture of poverty" from abject
generation to abject generation, it is a hopeful irony that the Jatavs
of Agra City, poor of the poor, urban and disadvantaged, have used
their own Indian customs and have created their own versions of
the Indian tradition, to adapt and to use, for themselves, new
institutions and new opportunities in coping with that very
poverty.

Conrad M. Arensberg

* *American Anthropologist,* Vol. 68 (No. 6), December, 1966.

ACKNOWLEDGMENTS

Life in a foreign land is often difficult and trying. My experience in India, although it had such moments, was quite to the contrary. The time that went into gathering the story of the Agra Jatavs was for me more than a research year abroad; it was a deeply personal and enriching experience. The people of Agra, especially the Jatavs, not only made this book possible, but they also made me a different and better person. Of the Jatavs I asked much and often; of me they asked only that I tell their story. While words are little thanks, I hope that the following pages in some way express my indebtedness to them.

I would like to thank each and every person in India who helped, encouraged, and befriended me. Though this is impossible, it would be ingratitude not to mention a few who profferred a warm and helping hand. Dr. M. N. Srinivas took time from an already impossible schedule to guide and advise me. Dr. Indera Paul Singh graciously opened his home and generously gave of his knowledge of Indian society. Were it not for the hospitality of Shri Ishwari Prasad Maurya, I might never have stayed in Agra. Ayodhya Prasad Kardam was a true friend whose honesty, integrity, and devotion to his people I could always rely upon. Master Man Singh, too, took an interest in this research and freely gave of his time and knowledge of Agra politics. The Mathur family helped me to understand and appreciate family life in India. If I have an adopted family in India, it certainly is the Pawars of village Kanapur in Madhya Pradesh. They nursed me in illness and regaled me in health. Shri Jahangir Singh Jayant helped me as an assistant and interpreter. In a sense, this book is as much his as it is mine. Shri Rajeshwar Prasad, Dr. Yogesh Atal, and Shri Prem Prakash Goyal of the Agra Institute of Social Sciences and Shri S. K. Khattri of the Delhi School of Economics, scholars all, provided food for both mind and body. A special word of thanks is due to my friends Carolyn Schaefer, Leona Chilemba, Don Rosenthal, Eleanor Zelliot, and Terry Heipp.

This book is a revised version of a Ph.D. dissertation presented to Columbia University in 1964. Special thanks are due, then, to the members of my dissertation committee. In India there is a tradition that a man ought to have a *guru* or a teacher who passes on to his disciple that special knowledge which cannot be found in books. Dr. Conrad M. Arensberg of Columbia University has been my *guru*. For more than seven years at Columbia, he has passed on to me the kind of anthropological culture that is the background and organization of this book. He is the gentleman described by Confucius (Analects XII, 16) who seeks to perfect the good qualities of men; he does not call attention to their defects.

A *padpadarshak* is a pathfinder or one who shows the way. Dr. Morton Klass of Barnard College is a *padpadarshak*. Although for some years now he has been a professor, he has not forgotten what it means to be a student. Were it not for such a friend my way to completing this work would have been full of discouraging false starts and time-consuming dead ends.

Dr. Morton Fried made me aware among other things of the importance of demographic data. Dr. Robert Murphy *malgré lui* has infected me with some of his own fascination with *l'anthropologie structurale*. Dr. Ainslie Embree has made me acutely conscious that India has a past as well as a present. I have also profited greatly from extended discussions with Dr. Herbert Passin, whose knowledge of Asia cannot be equaled.

Two others not on the dissertation committee have been of great help to me during my graduate training. Dr. Royal Weiler, now of the University of Pennsylvania, revealed to me the wonders of classical India and the pleasures of properly cooled *soma*. Dr. Harold Conklin, now of Yale University, helped make it possible for me to go to India, and he supported me in many other ways.

Many are my friends and relatives who in countless ways have helped to bring this work to completion. Betty Wood, Harriet Bloch, and Susan Lombardi translated my scribbled pages into clearly typed manuscript, while Sukey Waldron helped make sense out of my convoluted prose. One especially, J. M., I can thank only by asking her to share with me this accomplishment and all it means. For similar reasons I am indebted to Joe Potter, Dave Carney, and all my fellow students at Columbia University. Without such people the time and effort of graduate school might have weighed heavily.

The sixteen months of field work upon which this book is based were carried out from January 1963 to June 1964. Part of this research and two years of graduate study were supported by National Defense Language Fellowships. The greater part of the research and more than a year of writing time were supported by a research assistantship under project number MH 06227 of the National Institute of Mental Health. A State University of New York Research Foundation Faculty Fellowship was helpful in enabling me to revise and prepare the manuscript for publication.

I am grateful to the following authors and publishers for permission to use large quotations from their works: M. N. Srinivas, *Religion and Society Among the Coorgs of South India,* Asia Publishing House; Robert K. Merton, *Social Theory and Social Structure,* The Macmillan Company; F. G. Bailey, *Tribe, Caste and Nation,* the Manchester University Press; W. H. Morris-Jones, *The Government and Politics of India,* Hutchinson University Press; Pauline Mahar Kolenda, "Changing Caste Ideology in a North Indian Village," *The Journal of Social Issues* and "Religion, Caste and Family Structure: A Comparative Study of the Indian 'Joint' Family," in *Structure and Change in Indian Society,* Aldine Press; Marc Galanter, "Law and Caste in Modern India," *Asian Survey;* The Macmillan Company and Cohen and West (London), publishers of *The Theory of Social Structure* by S. F. Nadel.

Owen M. Lynch

CONTENTS

THE POLITICS OF UNTOUCHABILITY

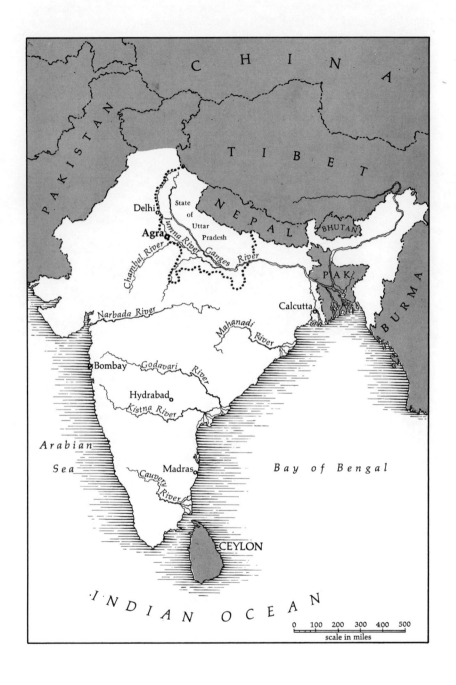

CHINA

PAKISTAN

TIBET

NEPAL

BHUTAN

Delhi

State
of
Uttar
Pradesh

Agra

Chambal River

Jumna River

Ganges River

PAK.

Calcutta

Narbada River

Mahanadi River

Bombay

Godavari River

Hydrabad

Kistna River

BURMA

Arabian
Sea

Madras

Cauvery River

Bay of Bengal

CEYLON

INDIAN OCEAN

0   100   200   300   400   500

scale in miles

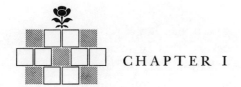

CHAPTER I

Introduction

About twenty years ago a noteworthy event signaled the beginning of a new era in world affairs. This was the achievement of independence and self-rule by the people of India. Thereafter one nation after another has achieved independence in a now postcolonial world.

While we know something about the broad political problems of the new independent nations, our information concerning the effects of independence upon various groups within these states is still minimal. This book is meant as a contribution toward filling this gap. It is particularly concerned with the effects of independence and of the adoption of parliamentary democracy as a system of government on one small group of people in India, the Jatav caste. More explicitly it is about one section of this caste, that which lives in the city of Agra, in the state of Uttar Pradesh, India.

This book is also meant as a contribution in another way. When anthropologists apply their methods and techniques to the non-Western world, they traditionally study either tribes or peasant villages. In this study of ("with" would be more appropriate) the Jatavs, I have applied anthropological techniques and methods to a non-Western urban scene; the Jatavs have a long history of residence in Agra city. Thus, the following pages are an attempt to show that anthropology has something valuable and valid to say about the urban scene as well as about the primitive tribal and rural peasant scenes. In short, this is a contribution to the burgeoning field of urban anthropology.

The Agra Jatavs are one of a much larger group of castes known as Camar, or leather workers (though some Jatavs would deny that they are Camars, as we shall see). In the traditional caste system of India Camars fall close to the bottom rung of the caste hierarchy because they work with leather, which is a polluting object, and because they are reputed to eat beef, which is the most polluting of foods according to orthodox Hinduism.

Ethnographic Problem[1]

In pre-independence India the Jatavs had begun a process of self-evaluation and self-reformation in an effort to gain respectability and higher status in the caste system. They gave up eating beef and tried to adopt many of the more orthodox Hindu rituals. However, in post-independence India they have rejected Hinduism and the caste system as well as the Congress Party, the now dominant, ruling, and most powerful party in India. Instead, they have become Buddhists and have formed their own political party, the Republican Party of India.

These dramatic changes in Jatav behavior are an ethnographic problem that require explanation. Why did the Jatavs change from an acceptance of Hinduism and the caste system to a rejection of them? What were the social and historical circumstances that surrounded and brought about this change? Briefly, the answer lies in the changes which have taken place in the structure of post-independence Indian society; changes which have created a new socio-political environment in which castes interact. These changes, such as the acceptance of the universal franchise, the adop-

[1] I am following Goodenough (1956) here in his distinction of ethnography, as the constructing of a theoretical model of a society, from ethnology, as doing comparative studies or the seeking of generalized or universal principles.

tion of a parliamentary system of government, an increased availability of education for all, the abolition of Untouchability, and the adoption of a planned system of economic growth, have created new potentialities for changed patterns of interaction between castes. The nature of some of these changes is so drastic and so extensive that there is little doubt that an alien government such as the British could ever have carried them out. They are the work of an indigenous elite who alone have any hope of bringing them to fullest fruition (Srinivas 1966: 89).

This new environment shows a caste, such as the Jatav caste, to be an "adaptive structure" (Gould 1963)[2] which can respond to the new means for social action provided in post-independence India and thereby gain a greater share of scarce goods such as power, prestige, wealth, and education for its members. In this study, the unit or population which is adapting is that group of individuals known as and belonging to the Jatav caste. Thus, our ethnographic hypothesis is that a caste is an "adaptive structure" through which its members can relate and re-relate themselves to members of other castes in terms of the potentialities provided to them by the socio-political environment in which they interact.

Ethnological Problems

The job of the social scientist is not only to describe in ethnographic detail the particular object of his study, but also to make comparisons of his data with those of others and to develop general principles and theories or frames of reference which organize and, hopefully, explain his data. In this book, then, I attempt to go a step beyond answering the ethnographic hypothesis and I consider certain ethnological problems which the Jatavs of Agra exemplify.

[2] "Caste is not disappearing in India because the solidarities inherent in it perform important functions in contemporary transitional society" (Gould 1963: 427).
". . . as India changed in response to the forces of modernization, certain castes must have possessed *qua castes* certain advantages or disadvantages vis-à-vis the opportunities for power, occupations, and social mobility which the new situation afforded" (Gould 1963: 430).

In the mind of the thinking layman and the professional social scientist alike, caste, as a system of social stratification,[3] has been considered the model of a static, rigid, and hierarchical social system. The Indian caste system is often taken as the prototype and as the most unchanging of all caste systems. Recent field studies, however, have begun to alter this view of the Indian caste system. They have shown it to be subject to dynamic change and internal group mobility. Barber (1968) has summarized these developments and shown how the old view of the Indian caste system was based on philological and textual materials. This textual view is giving way to a contextual view based on empirical case studies.[4] These new case studies often point out that there is a peculiar dynamic process of change in the Indian caste system known as Sanskritization. In this study, rather than accepting Sanskritization as an explanatory concept, I want to focus on it as problematic and to question its analytical utility and power.

The earliest explicit statement of Sanskritization was given by M. N. Srinivas (1965) in his book *Religion and Society Among the Coorgs of South India.* In that book he describes Sanskritization thus:

The caste system is far from a rigid system in which the position of each component caste is fixed for all time. Movement has always been possible, and especially so in the middle regions of the hierarchy. A low caste was able, in a generation or two, to rise to a higher position in the hierarchy by adopting vegetarianism and teetotalism, and by Sanskritizing its ritual and pantheon. In short, it took over, as far as possible, the customs, rites, and beliefs of the Brahmans, and the adoption of the Brahmanic way of life by a low caste seems to have been frequent, though theoretically forbidden. This process has been called "Sanskritization" in this book, in preference to "Brahmanization," as certain Vedic rites are confined to Brahmans and the two other "twice-born" castes (Srinivas 1965: 30).

[3] I distinguish caste as a unit in the system from caste as the system of stratification itself. The use of the same word for both referents has led to much confusion in the literature.

[4] It is rather interesting to note, however, that early nineteenth-century British administrators, such as Mountstuart Elphinstone, left realistic descriptions of mobility in the caste system. These seem to have been ignored by later sociologists.

In two later essays Srinivas (1956; 1957; reprinted in 1962) developed his earlier insights. In particular, he amended the view that it was only the Brahmans who were imitated in Sanskritization and said (1962: 43) that it might be the "dominant" caste of a region which was imitated rather than the Brahman caste. This amended point of view is reflected in his latest definition of the process: "Sanskritization is the process by which a 'low' Hindu caste, or tribal or other group, changes its customs, ritual, ideology, and way of life in the direction of a high, and frequently, 'twice-born' caste" (Srivinas 1966: 6).

Srinivas has also noted three conditions which qualify the concept of Sanskritization. First, it is a group process and is not applicable to individuals. An individual is always identified with a caste and Sanskritization must be valid for the whole caste in order for it to be valid for individuals. Second, it is a process that takes a number of generations before it is successfully accomplished. Third, it seems to be a process that has never worked for Untouchables who are below the ritual barrier of pollution.

The definition of Sanskritization as developed by Srinivas remains a problem for the social anthropologist, since it is a definition in cultural and not in social-structural terms.[5]

To describe the social changes occurring in modern India in terms of Sanskritization and Westernization is to describe it primarily in cultural and not structural terms. An analysis in terms of structure is much more difficult than an analysis in terms of culture (Srinivas 1962: 55).

Sanskritization is not, however, the only process of change in India. Srinivas (1962: 49–62) has noted that a process of Westernization began with the advent of the British. This involves the acceptance of Western dress, diet, manners, education, gadgets, sports, values, and so forth, though it seems to vary in detail from region to region in India (Srinivas 1962: 50–51).

The net result of the Westernization of the Brahmans was that they interposed themselves between the British and the rest of the native

[5] I take it that when Srinivas refers to structure in opposition to culture, he means social structure, since culture, too, can be structured.

population. The result was a new and secular caste system superimposed on the traditional system, in which the British, the new Kshatriyas, stood at the top while the Brahmans occupied the second position, and the others stood at the base of the pyramid (Srinivas 1962: 51).

The quotations above (from Srinivas 1962: 51 and 55) point out two important features of the Westernization concept of Srinivas. First, Westernization as well as Sanskritization is defined in cultural, not structural terms; second, Westernization, so defined, does not necessarily imply a structural change in Indian society. Note how Srinivas says that "the result was a new and secular caste system"; that is, it was not the structure of the caste system as such which changed, rather it was only the cultural symbols and style of life identifying those at the top of the hierarchy which changed. The British were "the new Kshatriyas," in the "pyramid" of the caste system. For this reason, I will not use the term Westernization when I refer to basic structural changes in post-independence Indian society. For the same reason, too, the discussion of Sanskritization in this study applies as well to Westernization, since Westernization implies only a change in the system of "symbolic justification" (Barber 1957: 404), and not in the system of social structure.

The above discussion of Srinivas's concepts leads directly to a set of three interrelated problems which I will treat in this study. The first problem is: Can Sanskritization be described in social-structural terms? Using case study materials from my study of the Jatav Camars of Agra, I will develop a frame of reference for the social-structural analysis of Sanskritization. As a corollary to this problem I will attempt to generalize this frame of reference so that it will be applicable to cross-cultural and cross-temporal (diachronic) analysis of other mobility movements. This corollary arises out of the fact that Sanskritization is a culture-bound concept and, as such, is useless for the comparative structural analysis of mobility movements.

The second problem to be investigated concerns the applicability of Sanskritization and Westernization to *all* attempts at caste mobility in India. Given the attainment of independence and the promulgation of a democratic constitution with a parliamentary form of government, it is not unreasonable to suspect that mobility in India might take some new form. This problem was first pointed out by Gould in 1961. He notes:

By the time they [the low castes] reach their destination [of Sanskritization], however, they will discover that the Brahman has himself vacated the spot and moved on to the higher hill of Westernization where he still gazes contemptuously down upon them from an elevated perch. . . . No doubt it will be at this point that the lower castes also commence abandoning their craze for Sanskritization and then the book will have to close on this concept, as the resultant new Indian society comes to grips with the problem of hierarchy in radically different and at this juncture hardly foreseeable terms (Gould 1961: 949).

Evidence will be presented which demonstrates that the Agra Jatavs have reached the point which Gould speaks of and that they are, as he also suggests, coping with the problem in radically new, though now foreseeable, ways.

Thus, the second problem is: Are the terms "Sanskritization" and "Westernization" applicable to all caste mobility movements? If not, then what is replacing them? My thesis is that political participation for some castes, such as the Jatavs, is replacing and is a functional alternative to Sanskritization and Westernization. This thesis is based on the fact that the political, and to some extent the economic and social, environment in which Indian castes interact has changed. The new constitution is based on parliamentary democracy and the universal franchise. Furthermore, India is trying to change its economic and social base through a planned program of development. Therefore, the *means* available to castes aspiring to be mobile are no longer limited to Sanskritization, though the *end* may remain the same, that is, higher social status. (I shall say more about *ends* and what they mean in the body of the book.)

If for the moment we can assume that the first two problems have been solved, one further problem still remains. The structural analysis of Sanskritization and its functional alternatives is, in effect, an analysis of caste external or caste-to-caste relationships within the caste system. Changes in the external relationships of a caste, such as the Jatavs experienced, can be expected to have a disturbing feedback upon the internal relationships of members of the caste with one another. Thus, the third problem is: How do changes in the external relationships of a caste affect its internal system of social organization? The methodological corollary of this problem is to ascertain whether the analytical frame of reference used in the analysis of the first two problems can be extended to include this third problem as well.

To sum up, then, the problems raised in this study are:

1. Can Sanskritization be defined in terms of social structure? And, as a corollary to this problem, if it can be defined structurally, can Sanskritization be fitted into a larger frame of reference, that is both cross-cultural and cross-temporal (diachronic)?

2. Is there a functional alternative to Sanskritization and Westernization in modern India?

3. Assuming a caste is Sanskritizing or has adopted some functional alternative to it, can a structural description of the effect of these caste external changes upon caste internal social organization be given? And, as a corollary to this problem, can a structural description of caste internal change be included in the larger frame of reference developed for the analysis of caste external change?

Analytical Concepts Used in Treating the Data

In order to make an analysis of the upwardly mobile Agra Jatavs, I have relied heavily on reference group theory as developed by Merton (1957). Applications of this theory in anthropology have already been succinctly summarized by Berreman (1964) and, therefore, need not be repeated here. Reference group theory is used in this study for the purpose of defining the social situation of a group aspiring to be mobile. In other words, it defines how it has sized up the situation. Once the situation has been defined, then organized action for social mobility can take place. Such action generally leads to conflict with other groups within the larger society. For the structural analysis of such conflict, I have relied on status-role theory as developed by Merton (1957) and Nadel (1957).

The sociological aspects of reference group theory have been summarized by Merton.

> That men act in a social frame of reference yielded by groups of which they are a part is a notion undoubtedly ancient and probably sound. Were this alone the concern of reference group theory, it would merely be a new term for an old focus in sociology which has always been centered on the group determination of behavior. There is, however, the further fact that men frequently orient themselves to groups *other than their own* in shaping their behavior and evaluations, and it is the problems centered about this fact of orientation to non-membership groups that constitute the distinctive concern of reference group theory. . . .
> In general, then, reference group theory aims to systematize the determinants and consequences of those processes of evaluation and self-appraisal in which the individual takes the values and standards of other individuals and groups as a comparative frame of reference (Merton 1957: 234).

In using this theory, the data have forced me to identify three, among many possible, types of reference groups. First, there is a reference group of *imitation* whose ways of behavior are accepted as right and proper or as useful and therefore to be imitated by the group making the reference.[6] Second, there is a reference group of *identification* to which an individual refers when identifying himself. He may do this when he is actually a member of that group or when he merely claims membership in such a group. The latter alternative is of distinctive concern in reference group theory. Finally, there is a *negative* reference group which stands as one's enemy or as the denier of the claims of one's own group. These three reference groups are analytical types and may be

---

[6] This is similar to Turner's (1956: 328) reference group of identification. But since the group imitated and the group identified with may be different, we have preferred the present nomenclature.

overlapping in a concrete case; that is, they all may be located in one concrete group, although this is not always so. For example, the Jatavs of Agra now identify with Buddhists throughout the world. However, they imitate the tactics of other Indian political parties such as the Congress and the Jan Sangh, and their negative reference groups are the orthodox Brahmans and the rich upper castes and classes. These three reference groups provide answers to three questions basic to the definition of the social situation of a socially mobile group. These are: (*a*) Who are we (or who do we claim to be)? (*b*) How must we behave in order to validate who we are (or claim to be)? (*c*) Who is blocking our way or rejecting our claim?

This theory is extremely useful to the social analyst because it enables him to identify the socially structured frame of reference in which a mobility aspiring group defines its situation. In terms of this definition of the situation, a mobile group decides what actions it will take to achieve its goals.

Reference group behavior occurs within a social system, and in this study the caste system is paramount. I consider caste to be an involute system, that is, a system based upon mutually exclusive status-sets or sub-sets. The most recent statement of this model is that of Bailey (1963), who builds upon the insights of Barth (1960), Leach (1960), and Nadel (1957). Bailey (1963) lists six criteria which define a caste system; they are: (*a*) Exclusiveness (membership in one group of the same type excludes membership in all other groups of the same type); (*b*) Exhaustiveness (there are no non-members in the society); and (*c*) Rank (groups are hierarchically ordered).

These first three criteria define any system of stratification; therefore, three additional criteria are added to define a caste system. It is: (*d*) Closed (recruitment to a group is ascribed by birth); (*e*) Involute (relations between groups are organized by role summation (cf. *infra*)); and (*f*) Cooperative (groups in the system do not compete).

To these six criteria I would make the following qualifications and clarifications. First, Bailey's criterion of cooperation[7] views a caste system as organic and static; that is, each group knows its place and keeps it, so to speak, by unfailing performance of its duties relative to all other groups in the system. Competitive relationships (politics for Bailey) take place only within the dominant caste of the hierarchy but not between castes in the system. Bailey's definition, it seems to me, is too rigid and does not allow for the dynamic processes and elasticity of actual contingencies within a caste system.[8] I believe there can be some forms of competition in a system so modeled. To allow for this, I have made ues of Bailey's own distinction between conflict and contradiction.

"Contradiction," as distinct from "conflict," is primarily an heuristic device used to diagnose the presence in a social situation (or social "field") of more than one structure. . . . Conflict appears at the level of dynamic analysis; and it can only be recognized as a contradiction by the absence of self-regulating factors. . . . If a man follows rule A and deviates from rule B, and if there is a third rule or institution designed to settle such situations on the grounds that *in this particular situation* one or the other rule is inappropriate, then this is not a contradiction . . . if the group which comes into action is not neutral between A and B but is in fact one of the sanctions of B (or A, as the case may be), and if it is effective, then there is a contradiction between the two allegiances and not merely a conflict (Bailey 1960: 239).

Thus, conflict can exist in the caste system when there is competition for higher rank through the self-regulating factor of Sanskritization. Sanskritization is the traditional means of mobility for a caste whose economic or other rank has improved and is therefore out of place with its low ritual rank; through Sanskritization ritual rank is also raised. It, thus, keeps the actual conditions of the Indian caste system in accord with the theoretical

[7] The criterion of cooperation, as Bailey has defined it, seems to me to be a euphemism to conceal a Radcliffe-Brownian universal functionalism. (See Merton 1957: 25–32.)

[8] Even in such an analytical model of a caste system such as Bailey's one can "build into" it dynamic processes so that it is more in accord with the concrete reality. This is not, however, to mix the two levels of the concrete and the analytic.

model of it. In other words it keeps the system in dynamic equilibrium. On final analysis, competition for higher rank through Sanskritization is really competition for those strategic resources to which higher caste rank gives legitimate access. This point is elaborated in the main body of the book.

Contradiction, on the other hand, is "symptomatic of social change" (Bailey 1960: 7), because it is evidence of the clash between two mutually incompatible systems, such as caste and democracy, within the same society. When contradiction exists, traditional means, such as Sanskritization, for resolving such situations will not ordinarily work.

A second clarification that can be made about Bailey's model of the caste system is that the criterion of involuteness might better be considered as the organization of relationships among groups through mutually exclusive status-sets or sub-sets.[9] This means that each caste is defined by a set, or at least a sub-set, of statuses which are idiosyncratic to it and which form the basis of its identification, ranking, and interaction within a caste system. The polar opposite of this is a hierarchy in which all statuses are open and cross-cutting, as in a class system.

A third clarification is that these criteria for a caste system form an abstract, analytic, and comparative structural model of the system; the model does not, then, give an exact replica of any particular caste system in reality. The difference between a real class and a real caste system is based upon which end of the continuum, from mutually exclusive to cross-cutting status-sets, they approach. In this study, I will use the terms "mutually exclusive" and "cross-cutting" in preference to "involute" and "non-involute."

A final clarification is that since Bailey's model of a caste system is abstract and social-structural, it necessarily prescinds from the cultural aspects of a caste system. In India the caste

[9] For the ideas of mutually exclusive and cross-cutting status-sets, as well as the ideas of "dominant," "controlling," and "salient" status, I am indebted to the stimulating lectures of Dr. Robert Merton. Any misunderstanding or misuse of these terms is, however, my own responsibility.

system is manifested in its own unique cultural idiom; that is, in terms of ritual pollution. Thus, a Brahman is purer than a Kshatriya and a Kshatriya is purer than an Untouchable. It follows that an Untouchable is polluting for the Kshatriya and more polluting for the Brahman. To avoid pollution from another caste, intercaste marriage and intercaste dining are forbidden. It is ritually polluting for a higher caste member to have sexual relations with or to take food from a member of a lower caste. The importance of knowing this idiom is comparable to knowing the language of one's own country; one must speak the language of his country in order to communicate within it. So too, one must know the idiom of his country's social system in order to interact within it. Thus, if an Indian wants to rise in the Indian caste system he must know and manipulate the idiom of ritual pollution. It would be useless for him to use the idiom of race which is used in the American caste system. Yet, though languages may differ, they are capable of conveying the same meanings. So too, though the caste systems of India and the United States may differ in their cultural idiom or manifestation, their basic structure or grammar may be similar or identical and they, thus, may have the same structural results for those within the system. Bailey's model of the caste system is most suited to these underlying structural problems, and for this reason I have adopted it in this study.

A number of other concepts used in this analysis remain to be defined. The first is the notion of "dominant" status. A dominant status is one which Ego asserts ought to take precedence over all his other statuses in a particular interaction with Alter. The second notion is that of "salient" status, which is that one of a number of possible statuses, other than his asserted dominant status, that Alter imputes to Ego. When Alter imputes a salient status to Ego, then either conflict or contradiction, as I have defined them, is present. For example, an Untouchable (Ego) asserts that he is a Kshatriya. A Braham (Alter) makes salient Ego's Untouchable status and thereby fails to accord him the Kshatriya dominant status. This is a situation of conflict since neither the Brahman nor the Untouchable is questioning the legitimacy of caste statuses as such. It is only a question of whether or not the Untouchable may occupy or be accorded the Kshatriya status he is trying to claim. However, when an Untouchable claims a dominant status of citizen and the Brahman accords him the salient status of Untouchable, then it is a case of contradiction. It involves a rejection of an asserted dominant status by the one and a rejection of an imputed salient status by the other. It also involves a rejection of one or the other social systems underlying these statuses. It is, in effect, an attempt of the Untouchable not only to assert his citizenship status, but also to compel the Brahman to act as a citizen too. A Brahman can interact with both a Kshatriya and an Untouchable, but he cannot interact with a citizen since neither Brahman nor citizen is a counter status of the other. Brahmans can only interact with other castes in the caste system; citizens can only interact with other citizens in a democratic system.

Another notion to be defined is that of central or controlling status.[10] A "controlling" status is one which normatively limits the combination of statuses in a status-set. In other words, it is the status one ought to occupy as a precondition for occupying certain other statuses. One who occupies a status without first occupying its controlling status contradicts the norms of his

[10] Nadel (1957: 31–41) would consider this a "basic" or "pivotal" attribute of a role (read status). "In normally entailing all other attributes [read statuses] it also legitimizes them, so that, in its absence, the other attributes appear as illegitimate, unexplained, or with an altogether different meaning" (Nadel 1957: 32). This basic or pivotal attribute, according to him, belongs to the "prehistory of a role." Thus, when he develops his typology of roles into "recruitment" and "achievement" roles, he concentrates not on the basic or pivotal attribute but on the roles resulting from that attribute. For example, a shoe maker in India is a "recruitment" role, since its basic or pivotal attribute is that one is born as a member of the Camar caste, which is again part of the prehistory of the shoe maker role (or status). However, this basic or pivotal attribute is not always an attribute but rather it is often a role (status) in its own right. Such a status is, therefore, a "controlling" or "central" status in my terms. Once knowing the "controlling" status, one can to some extent predict the statuses that can be expected to be correlated with it.

It seems important to note here the similarity in Merton and Nadel's thought. Though they concentrate on different aspects of a status-set, they seem to have independently arrived at similar insights into the nature of status.

society and is thus generally subject to negative sanctions. For example, in India one ought to be a Camar before he is a shoe maker. Not all Camars are shoe makers, but if one is born a Camar one can occupy the status of shoe maker without sanction and without breaking caste rules.

These are analytic concepts and therefore may overlap in a concrete case. However, the distinction is clear. Dominant and salient statuses are concepts used to analyze actual interactions, while controlling status is a concept used to analyze the rules governing the combination of statuses in a particular social structure.

Another concept yet to be defined is that of structural "observability." This

is conceived as a property of groups. It directs attention to the ways in which the structure of the group affects the input of information and the output (of response) which thereupon works to exert social control. . . .

It is a name for the extent to which the structure of a social organization provides occasion to those variously located in that structure to perceive the norms obtaining in the organization and the character of role-performance by those manning the organization. It refers to an attribute of social structure, not to the perceptions which individuals *happen* to have (Merton 1957: 321 and 350).

Structural observability can be somewhat better understood by distinguishing it from structural visibility. "Visibility" is that characteristic of a status which makes it known through the operation of status cues or attributes.[11] Thus, a soldier's uniform or a wedding ring make it apparent to those who can see and know the meaning of these cues that a person occupies the status of soldier or married person, respectively. The calluses on a shoe maker's hands make his occupational status visible to those familiar with the social meaning of such calluses. In a shoe factory, the shoe makers are visible to the manager of the factory because of the workers' calluses. They are also visible in the same sense to

---

[11] Nadel (1951: 67; 1957: 34) would call them diacritical signs.

the owner of many factories because he, too, can recognize the status cue implied in the calluses.

But there is an important difference between the manager of one factory and the owner of many, and this difference is a consequence of the different positions they hold in the structure of the entire shoe company. The manager has much greater first hand knowledge of the workers and their role performance than does the owner. Because of his position at the top of the organizational structure of the business, the owner has less observability **and,** therefore, less knowledge of the role performance of the workers in his factories than does the manager at the head of each local factory. Greater or less information gained as a result of the observability inherent in certain statuses *may* also result in greater or less control. Thus, while the owner has greater power and authority in the business, nevertheless he has less direct *control* over his workers' role performance than does the manager of the local factory. The proverb "while the cat's away, the mice will play" makes indirect animadversion to this variable property of statuses in a social system.

Finally there is the notion of "dichotomization," an infelicitous name for a particular type of social integration. The name arises out of the analytic distinction Nadel made between the correlative and autonomous aspects of a status. However, the concept actually refers to one of the two processes[12] of social integration recognized by Nadel. He has defined this important concept in the following ways:

The correlative aspect of any role [read status] (of whatever degree) relates it to a limited range of persons—the persons in counterpart roles—towards whom the actor exercises the rights and obligations intrinsic to the role; while its autonomous aspects render it capable of being related beyond that range, . . . to a wider (or widest) "public." No social system leaves the latter possibility unutilized. "Fathers" (as men having children) or "wives" (as women having husbands) will

[12] The other process is involution or the organization of a society through mutually exclusive status-sets. (See pp. 10 and 12.)

assume duties and entitlements valid in the society at large, over and above those they have towards their children and husbands. . . .

For what happens here is that the dyadic relationship of persons in correlative roles comes to involve the interests and reactions of "third parties." It may do so, quite simply, because the dyadic relationship is concerned with providing goods or services needed by a "third party," with managing a state of affairs affecting the interests of the latter or, more simply still, because it involves the cooperation of two persons vis-à-vis a third (or a plurality of "third parties") (Nadel 1957: 85–86).

Dichotomization, then, is concerned with dyadic relationships of a status and a counter status but analytically specifies what happens to the dyad when a third party enters into the relationship. The intrusion of a third party changes or influences the roles of the parties in a dyad. A special case of dichotomization is "politicization" in which the role performance in certain statuses becomes the concern of political authorities (Nadel 1957: 88).

The process of dichotomization is theoretically interesting for two reasons. First, its connection with the concepts of conflict and contradiction, as I have defined them, is obvious. It is conflict or contradiction which often brings a third party into the social situation. In the case of conflict the third party exerts its authority to settle the dispute in a traditional way without changing the social system; it operates to preserve the social status quo. In this sense dichotomization is merely a generalizing concept for various processes of social control. Were this its only meaning, the concept would be of little interest. When applied to situations of contradiction, however, dichotomization becomes of great interest and analytical utility.

In a case of contradiction, the third party exerts its power to change the dyadic relationships through redefinition of their roles toward one another and toward itself. Generally this process of redefinition is referred to as the *differentiation* of statuses. But note that when the third party enters, differentiation also involves an added relationship to the third party through the *integration* of newly differentiated statuses into some higher level institution.

Theoretically, then, dichotomization is an integrative process which links differentiation of statuses with their integration into higher level institutions under conditions of social change.

It is in this second sense that I shall use the concept of dichotomization in this book. In so doing I am conscious of the fact that I have interpreted and used the concept in a particular way. My justification for this, however, lies in two facts. First, I have used it to refer to a particular process of status integration as was Nadel's original intent. Second, Nadel's definition of the term is rather loose, unspecific, and ambiguous. I have merely tried to make the term specific and applicable to certain social situations.

The case of the Scheduled Castes in India offers a good example of dichotomization. The status of Untouchable vis-à-vis non-Untouchable has become the concern of the Indian government (here the third party). The government has redefined the status of Untouchable into that of Scheduled Caste and implicitly all other castes become non-Scheduled Caste; for this reason we can speak of status differentiation. Outside of this area of governmental concern the old caste statuses of the Untouchables and non-Untouchables, for all practical purposes, remain. In creating this differentiation of statuses the government has also brought those who occupy these new statuses into a special relationship to itself through integrating institutions at a higher level than caste. These new institutions are the Harijan Welfare Office and the Scheduled Castes Commission, whose duties include investigating cases of Untouchability and bringing such cases to Court. The Scheduled Castes are also being integrated into the other institutions of the state through the government's "protective discrimination" policy.

The second reason that dichotomization is of interest is because it has particular applicability to nations such as India, where concerted efforts at planned social change are being made. What this means is that new integrating institutions are being created by governmental fiat. These new institutions, therefore, stimulate from above status differentiation from below by requiring new credentials and role definitions for those who are integrated into them. The Block Development Plans and Panchayati Raj[13] are cases in point. Thus, change is taking place not just from below as a "natural" process of growth of more complex institutions out of less complex ones, but it is also being stimulated from above in a planned attempt to create a more complex, and presumably a more developed, society. In the following chapters specific examples of dichotomization are given, and its importance becomes clearer.

Conventions

Some points of clarification must also be made here about conventions used in the presentation of the data. Occasionally I refer to a "respondent group." This refers to the fact that I present some data in a quantified form which is not a sample in the technical sense. A "respondent group" is merely a number of arbitrarily chosen persons to whom I administered a schedule of questions to back up my own observations. In the text, the signs [ ] are used to indicate my own interpolations, while ( ) are used for interpolations already in a text. Throughout the text I use the spelling "Camar" rather than "Chamar" as often appears in other books.

I have adopted the convention of giving quotations from my own notes as though they were the words of my informants, when they actually are my transcription of what an informant said to me in the field. I have used this device in the hope that it will add some life and color to a story that might otherwise appear dry and lifeless. Moreover, the body of the text is written in the ethnographic present; that is, the book describes the Agra Jatavs as I had come to know them as of June 30, 1964, when I left India. While many changes have taken place in Agra since that time, they have not significantly affected either the analysis or the conclusions presented in this book. Finally, with three exceptions, I have used pseudonyms for all people and places in the book.

[13] Panchayati Raj is a government sponsored and initiated system of village governance. It is meant to be democratic and to encourage grassroots initiative.

CHAPTER II

Agra City and the Jatavs

History of Agra

Most visitors to India make a stop in Agra City to see one of the world's architectural masterpieces, the Taj Mahal. The trip is easily made because Agra is located in the southwest corner of the state of Uttar Pradesh about 120 miles south of New Delhi, the capital of India. Linguistically, Agra falls into that area of north India where a dialect of Hindi called Braj Bhasha is spoken, although one may also hear Hindustani and Urdu as he walks about the crowded streets and dusty markets of the city. The parched fields surrounding Agra mutely tell of an average annual rainfall of about 26.7 inches and a temperature range which rises to a scorching high of 41.5 degrees Centigrade in May and June and drops to a frosty low of 5.9 degrees Centigrade in January.

The events which form the historical substance of this book are but minor vignettes in the magnificent old tapestry depicting the history of Agra. The origins of the city are lost in antiquity and myth. The first historical mention of the city is in an account of its invasion by Sultan Ibrahim Ghazvani in about 1080 A.D. (Bhanu 1957: 2). Agra was of no real importance until 1502, when Sikandar Lodi moved the capital of his kingdom from Delhi to

Agra. Sikandar's successor surrendered Agra into the hands of the Emperor Babur after the battle of Panipat in 1526.

Under Babur, Agra City became the capital of the Moghul Empire in India. The Moghul emperors nurtured the city until it became the crossroads of their empire. In the last decades of the sixteenth century, the Moghul Emperor Akbar moved the city from the left bank of the river Jumna to its right bank where the city stands today. Akbar built the Agra Fort and in 1653 his grandson, Shah Jahan, built the famous Taj Mahal. Both of these buildings stand today and, because they attract tourists from all over the world, they add a cosmopolitan flavor to the city's life.

In 1658, during the reign of the Emperor Aurangzeb, the capital of the Empire was moved from Agra to Delhi. Even so, Agra continued to be of strategic and administrative importance. After the death of Aurangzeb in 1707, the Moghul Empire began to crumble under pressure of the internecine conflicts of his successors and the conquering advance of the Marathas. For some time the city was tossed back and forth among the Marathas, the Jats, and the Moghul Emperors. Not until 1803, when the British captured the city was a period of relative calm and prosperity restored.

The British in 1835 made Agra the capital of what was then the North-Western Provinces (Latif 1896: 65). It remained so until 1868 when the capital was shifted to Allahabad and Agra was reduced to the status of district headquarters. Except for the Sepoy Mutiny of 1857–58, which for the most part skirted Agra, the city under the British grew peacefully. Schools, Agra University, a hospital and medical school, railroad stations, a telegraph office, and ginning and spinning mills were all established. In the immediate post-mutiny era, Agra reached a climax in its development when the first Industrial Exhibition of 1867 was held. At the Exhibition "the manufactured industries and natural products of the District [Agra] were largely displayed" (Latif 1896: 67).

Near the turn of the last century, Agra's industry began to expand. The cottage shoe industry began, and has grown to the point that in India, Agra is second only to Kanpur City in the production of shoes. At present there are four cotton spinning mills, fifty-eight pulse mills of various sizes, and a number of oil pressing mills. There is also a small iron industry in the city and other small-scale industries are growing in importance (Indian Political Science Conference 1963: 10 and 50–51). However, Agra is not truly an industrial city, if "industrial" is defined as the use of large-scale power and machines (see Lynch 1963).

In 1942, during World War II, an airfield was built at Agra by the Third Air Depot of the U.S. Army. This group left in May, 1946 (Nigh 1965), but the airfield remains as a major depot for the Indian Air Force today. So too, an old British Cantonment (military base) remains as a major depot for the Indian Army.

Just as the city's industries and communication facilities have grown since the turn of the century, so, too, has its population. In 1901 the total population was 188,022 individuals, while today the total population of Agra City is 462,020 (India. Census Commission 1964: 206). In terms of population, Agra is now thirteenth among the cities of India and third among the cities of Uttar Pradesh (India. Census Commission 1953: 158).

The Agra of today is no longer the capital of an empire; it is only a district headquarters, a tourist center, and a center for cottage industries. Yet signs of a new life and a new spirit of independence can be sensed by anyone who cares to walk about and listen. What these changes are and how they are affecting one caste in Agra are questions I shall attempt to answer in this book. Now that we have set the scene, let us turn to the actors and ask, "Who are the Agra Jatavs?"

The Agra Jatavs

The Jatavs of Agra are a low caste, Untouchable community whose traditional occupation is leather working. They are also part of two larger groups known as the Camars and the Scheduled Castes. For reasons of clarity, then, I will proceed by answering the following questions: Who are the Scheduled Castes? Who are the Camars? Who are the Jatavs?

Within the caste system the Jatavs are Untouchables. That is to say, they are at the bottom of the caste hierarchy. They are polluting to the upper castes and therefore are the objects of discrimination; and, because of this low caste rank, they have remained, on the whole, illiterate, poor, and virtually powerless. Nowadays the Untouchables are called one of the "Scheduled Castes" [1] (S.C.) because of a special relationship they have with the government. A Scheduled Caste is one whose name appears on a list first issued by the British in 1935. The list remains substantially the same today as it was then and can be changed only by the direction of the President of India himself. Castes were placed on the list on the basis of an all-India set of criteria. These criteria included: exclusion from entry into Hindu temples; exclusion from the services of "clean Brahmans"; exclusion from the services of the same barbers, tailors, and so forth, used by higher castes; inability to give water to higher castes; and limited access to public facilities such as wells, schools, and roads. Where these criteria were not met, as in South India, then two other criteria were added: illiteracy and poverty. The purpose of these criteria was to identify those castes in India which were so low in the caste hierarchy that they were considered Untouchable, though often only figuratively so.

The purpose of the list was to identify those eligible for the benefits of "protective discrimination" (Galanter 1963: 551). In other words, the government would treat those castes on the list with special and unequal treatment vis-à-vis other castes. The reasoning behind the government's policy of "protective discrimination" was and is that great, and undesirable, inequalities exist between the Scheduled Castes and the non-Scheduled Castes; by temporarily creating conditions of unequal and favorable treatment for the Scheduled Castes, the government hopes eventually to achieve conditions of equality in the society at large. It was

---

[1] Unless otherwise stated, material for this section is substantially from Dushkin (1957; 1961a, b, c). But see also Galanter (1961; 1963) and the various Reports of the Commissioner for the Scheduled Castes and for the Scheduled Tribes published by the Manager of Publications, Government of India, Delhi.

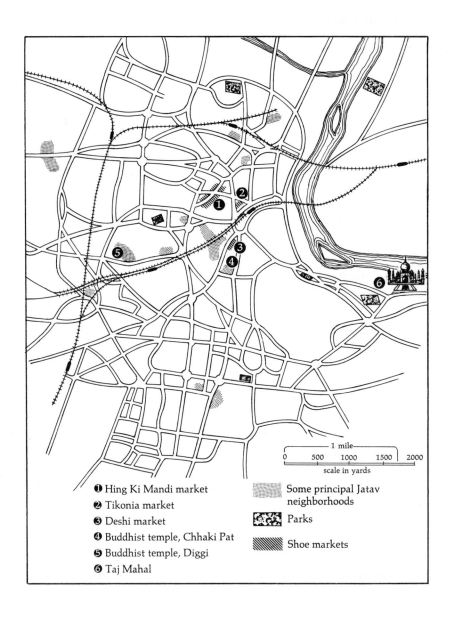

❶ Hing Ki Mandi market
❷ Tikonia market
❸ Deshi market
❹ Buddhist temple, Chhaki Pat
❺ Buddhist temple, Diggi
❻ Taj Mahal

Some principal Jatav neighborhoods
Parks
Shoe markets

1 mile
0   500   1000   1500   2000
scale in yards

assumed that this legislation was enacted for the elimination of social, economic, and educational inequalities. It was also assumed that, because of the close correlation of low social, economic, and educational status with Untouchable status, the elimination of the former would tend to eliminate the latter. In independent India, the legislation passed on this subject under the provisions of the Constitution of 1950 was originally to be in effect for ten years, from 1950 to 1959. However, it was extended for another ten years and is expected to end by 1970.

The provisions for "protective discrimination" might be grouped under three headings: education, government jobs, and political representation. In the educational sphere, the governments of the center[2] and the states have various schemes for providing scholarships, fee remissions, books, boarding grants, and so forth, to members of the Scheduled Castes. These privileges run from first grade to post-graduate studies and even include a few overseas scholarships.

In the sphere of government jobs, a certain percentage of jobs at the central and the state levels are reserved for members of the Scheduled Castes. At the center 12.5 per cent of posts recruited by examination and 16.7 per cent of posts recruited by appointment are reserved for the S.C.'s. In the states, reserved government jobs are equal to the percentage of S.C.'s in the total population of the state. In the state of Uttar Pradesh, this is 18 per cent.

In the sphere of political representation, "protective discrimination" means that there are a certain number of seats reserved for the S.C.'s in the People's and States' Legislatures (Lok and Vidhan Sabhas) at the center and the state level respectively. These are established according to the percentages of S.C.'s in the total population and are filled by joint electorates in which only S.C.'s may be candidates for the seat to be filled. Nevertheless both S.C.'s and non-S.C.'s vote for the S.C. candidate of their choice. However, any S.C., if he so chooses, may stand for a non-reserved seat, called a general seat. In this way he competes on a par with all

[2] The equivalent of the United States federal government.

other candidates, be they S.C. or not. Until 1962, reserved seats were only in double member constituencies, that is, certain constituencies had both a general and a reserved seat. After 1962, double member constituencies were abolished and divided into two constituencies, one general and one reserved. That half of the old constituency which had the greatest number of S.C.'s was given the reserved seat. (However, this law was not yet in effect in Agra by the time of the 1962 elections, which fall into the period covered by this study.)

In addition to the "big three" of education, government jobs, and political representation, the central and state governments sponsor welfare programs for the S.C.'s and campaigns for the eradication of Untouchability. The Constitution of India has, furthermore, abolished Untouchability and enacted many provisions against discrimination and inequality of treatment before the law.[3]

[3] "The important articles of the Constitution are:
(Fundamental Rights)
14—equality before the law
15—no caste discrimination by government or by private persons in regard to use of public facilities; special provisions in favor of untouchables and 'backward classes' permissible.
16—no caste discrimination in government service; reservation of places permitted for untouchables and 'backward classes.'
17—untouchability abolished and enforcement an offense.
23—forced labor abolished; no caste discrimination in regard to compulsory public service.
25 (2) b—freedom of religion qualified to save temple-entry laws and state power to legislate for social welfare and reform.
29—no caste discrimination in admission to state-aided educational institutions.

(Directive Principles)
44—personal law to be replaced by 'uniform civil code.'
46—State shall protect and promote interests of 'weaker sections of the people,' especially untouchables.

(Other Provisions)
325—no caste electorates.
330, 332—reservation of seats for Scheduled Castes in Parliament and State Legislatures.
334—ten-year limit on latter, extended another ten years by the Constitution (Eighth Amendment) Act, 1959.
335—claims of untouchables to be considered in appointments to government service.

In 1955 the Untouchability Offenses Act (UOA) was passed. It

. . . outlaws the imposition of disabilities "on grounds of untouchability" in virtually all fields of activity excepting home life, private religious ceremonies, and private employment. . . . The UOA punished not only the direct enforcement of disabilities but also indirect social support of untouchability (Galanter 1963: 550).

The term "Untouchability," however, is a slippery one and has been legally defined by the High Court.

In the words of the First High Court which considered the questions, it [Untouchability] includes only practices directed at "those regarded as 'untouchables' in the course of historic development"—i.e., those relegated "beyond the pale of the caste system on grounds of birth in a particular class" (Galanter 1963: 551).

The UOA is unique in that it presumes that a non-S.C. who has committed an offense under this Act is guilty until proven innocent.

The most unfortunate aspect of the government's policy of "protective discrimination" is its failure to undertake studies of the policy's effectiveness. The government knows everything about the goals and institutions it has set up under this program. Yet it knows practically nothing about the effectiveness of the policy itself. There have been no evaluative studies of the program except, it seems, one attempt by the Commissioner for Scheduled Tribes and Scheduled Castes. In 1961 the Commissioner sent an inquiry to all the state governments asking for information along these lines, but as of 1963 he had received not a single reply (Isaacs 1965: 19). The bits and pieces of information and impressions available lead to the conclusion that the various programs of the "protective discrimination" policy have been differentially effective. A hard judgment about how effective they are is difficult to make.[4]

338—special officer to investigate and report on safeguards to Scheduled Castes.
340—provision for appointment of a backward classes commission.
341—provision for specification of Scheduled Castes" (Galanter 1963: 549).
[4] Harold Isaacs (1965) in his book, *India's Ex-Untouchables,* gives the closest

In the sphere of political representation, S.C.'s are elected to all the reserved seats, but whether or not they forcefully and adequately represent the interests of those who elect them is debatable. Some progress has been made in the sphere of education; in fact, it seems to be the most successful part of the policy. In the sphere of reserved jobs and government services, however, "the position as regards the actual representation of Scheduled Tribes and Scheduled Castes in posts under the various Ministries of the Government of India and their Attached and Subordinate offices is not satisfactory" (India. Commissioner for Scheduled Castes and Scheduled Tribes 1962: 278). Such results seem to reflect adversely on the provisions for education, since those educated under the provisions must be either too poorly educated to get into government services or insufficiently educated to get into the higher posts. The effectiveness of the legal abolishment of the practice of untouchability and other forms of discrimination is meagre, especially in the villages. In a circular issued to all District Magistrates by the government of the state of Uttar Pradesh, it was noted that: "The practice of untouchability continues unabated. . . . Such occurrences bring a bad name to the Government and show that the provisions of Untouchability (Offenses) Act, 1955, are being disregarded on a large scale" (quoted in: India. Commissioner for Scheduled Castes and Scheduled Tribes 1962: 25).

Prominent among those known as Scheduled Castes is a large group of castes known as Camars. The word Camar is derived from the Sanskrit *charma kara,* which means leather worker. In earliest (Rig Vedic) times, the leather worker does not seem to have been an object of opprobrium and, indeed, his services were of utmost importance to the warriors of those days. In the ancient Indian epic, the Mahabharata, the Camar was the maker of shields, breast

---

thing there is to an evaluation of the effectiveness of the whole program. His book is based on a number of interviews with members of the Scheduled Castes, especially the Mahars of Maharashtra. It is well worth reading for the "feel" of what the program is doing for some individuals.

plates, and body armor, as well as of drums and various parts of chariots (Briggs 1920: 13).

Many theories have been put forth to explain the origin of the Camars; none has been established. It has been suggested that they are the product of mixed caste marriages, or that the unpleasant and evil-smelling occupation of leather tanning relegated them to the peripheries of respectable society, or even that they are part of a race conquered by invading Aryans. At most, there seems reason to believe that the Camars are a heterogenous group of peoples who have received from time to time recruits from castes higher up in the caste hierarchy (Briggs 1920: 17).

Mythical accounts of Camar origins match the ocean sands for number and variety. For example, there is the legend of the five brothers. While on a walk, five Brahman brothers came upon the carcass of a dead cow. Four of them walked on, but the fifth stopped and pulled the carcass off the road. For this act, his brothers excommunicated him and ever after it was his lot and that of his descendants, the Camars, to remove polluted and polluting dead cattle (Crooke 1896: 170).

The position of the Camars in India is very low indeed. In the villages they work as general menials for some traditional payment in grain, clothes, and food. They were, and in some places still are, subject to forced labor (*begar*); they own little land; and until recently have had little hope of improving their impoverished condition. The traditional occupations assigned to them are tanning hides, making shoes, and removing the carcasses of dead animals, which they eat. Since leather is a polluting object and beef is the most polluting of foods, the Camar who engages in such acts, and by association all other Camars, becomes polluted. Because of this pollution Camars, more often than not, are forced to live on the periphery of a village or in separate hamlets. Lillingston (1913: 351) quotes a proverb about the Camars, who are "supposed" to be dark in complexion, while the upper castes are "supposed" to be fair in complexion.

Karia Brahman, gora Chamar,
Inke sath na utariye par.

If the Brahman be black, if the Chamar be fair,
Let the wise beware, if cross the river he dare.

The meaning is that a fair-skinned Camar is such a rarity that
something *must* be wrong and an upper caste person ought to be
on his guard. For the Jatavs, such a proverb is, in the eyes of this
writer, no more than an unfounded stereotype.

As a group, the Camars are most numerous in North India,
especially in the states of Uttar Pradesh, Punjab, and Bihar. The
Census of 1901 showed them to be numerically the largest group
of castes in the former North-Western Provinces and Oudh, while
the Brahmans ranked second (India. Census Commission 1902:
180–229). There are numerous castes (*jati*) of Camars; the Census
of 1891 listed 1,156 of them (Briggs 1920: 21). The Jatavs are one
of these castes or endogamous groups of Camars, but they tend to
consider themselves separate from and higher than other Camar
castes.

Jatavs are found largely in the western part of Uttar Pradesh
State and northward up into Punjab State. They are sometimes
known as Jatiya, Jatua, or Jadav, but the name Jatav is current in
Agra City. "Some say that their name is derived from the word
jat, meaning camel driver; others, that their name connects them
with the Jat caste. It is sometimes said that they are descendants
from the marriage of Jats with Chamars" (Briggs 1920: 23; see also
Crooke 1896: 173; and Goyal 1961: 44).

Lillingston (1913: 352) notes that the Jatavs of Gorakpur District are served by Gaur Brahmans for priestly services and this,
therefore, gives them a high status. But this does not seem to have
been true for Jatavs in Agra. Briggs (1920: 22) writes, "The Jatiya
can reasonably claim to be the highest of all the sub-castes of
Camars." In a preliminary tabulation of the 1961 Census given to

me by the Census Office of Uttar Pradesh at Lucknow, there were
in Agra City Corporation 71,404 Camars, of whom almost all are
Jatavs. This figure can be reliably used for the total Jatav population, since it presumably does not include the 2,262 Buddhists,
almost to a man Jatav converts (India. Census Commission 1963:
40), who cannot be listed as Scheduled Castes.[5] The Buddhists,
therefore, more than substitute for any other Camar *jatis* included
in the total enumeration of Camars. The total population of Agra
City, according to the 1961 Census, was 462,020 (India. Census
Commission 1964: 206). The Jatavs, thus, constitute about 16
per cent, or one-sixth, of the city's total population. Another
preliminary tabulation from the Uttar Pradesh Census Office lists
the total Scheduled Caste population of Agra City as 97,833 individuals, making the Jatavs about 77 per cent of the total Scheduled Caste population in the city.

One other term often applied to the Jatavs must be mentioned
here, for I shall not allude to it again. The reasons are obvious,
but not always on the surface. As members of the Scheduled Castes,
the Jatavs are often called *Harijans,* a name given to all Untouchables by Gandhi. Literally the word means "child of God," but
figuratively its connotations are quite different. My Jatav informants showed a dislike—at times an intense dislike—for the
word. They felt it connoted the idea of being a bastard and also
brought to mind patronizing upper caste benevolence.[6] For these
reasons, I, too, avoid use of the term in this book.

[5] A Scheduled Caste by definition must be a Hindu in the State of Uttar Pradesh.
This is a problem for the Jatavs and it is treated in Chapter V of this book.

[6] Isaacs (1965: 166) specifically notes that only very few of the Scheduled Caste
members whom he interviewed made free use of the term.

CHAPTER III

The Sociology of the Market

Agra City is famous in India today as much for its ethereal Taj Mahal as for its pedestrian shoe industry. Most Agra Jatavs are engaged in making shoes. Things were not always so, for only in the present century has shoe manufacturing become a major industry of the city. How has this come about? And what relation does the Jatavs' place in the shoe market system have to the politics of Untouchability? These are the two basic questions of this chapter.

History and Structure

Before the growth of the shoe industry, Jatavs in Agra City were mainly stone cutters, laborers, scavengers, and tanners of leather.[1] All of these are occupations which they traditionally performed as part of the village economy. In that economy the Camars form a large part of the landless laboring class, and working in leather is only a subsidiary, although hereditarily ascribed occupation. Because of the pollution barrier, these demeaning occupations were

[1] The Census of 1881 lists 1,570 males engaged as shoemakers, sellers, hide dealers, tanners, and leather dyers (India. Census Commission, 1882: 194). However, the Census of 1911 lists 993 as dealers in hides and skins and 3,569 as boot and sandle makers, for a total of 4,562 (India. Census Commission, 1912: 345).

left to the Jatavs. It must be remembered that in Agra district alone, Camars in 1901 were 19.13 per cent of the total population (Neville 1921: 75). Some Jatavs became contractors for the labor of their own caste mates in the stone cutting, building, and bone meal industries. Notable among these early contractors were the two rich men (*seths*), Seths Sita Ram and Man Singh, who succeeded in opening a number of large cotton mills before their deaths in the 1890s.

After the turn of the last century the shoe industry in Agra began to grow, and it expanded greatly during World Wars I and II. By the end of World War II, markets for Agra shoes existed throughout India as well as in Iraq, Iran, and the East Indies (Sharma 1958: 59). These foreign markets were greatly reduced after the partition of India, since they were run by Muslims who had fled to Pakistan. However, after 1955, new orders for shoes were received from Russia and other Communist countries, as well as from Germany.

Since 1955, the volume of business with Russia and other Eastern European countries has continued to grow. This increased international trade has had two effects on the shoe industry. First, it has greatly sharpened the skills of the workers engaged in production for the overseas market who now produce fashionable, high-heeled leather shoes of a quality found in New York shops. Second, it has made the Jatav cottage workers very conscious of their great dependence upon international trade. In early 1963, a knowledgeable informant estimated that about 13 per cent of the present shoe trade in Agra was with Russia and its satellite countries. If this is so, and if, as seems likely, such foreign orders have increased, the results of a cut-back could be disastrous for the Jatav workers who are directly dependent upon exports.

As the shoe making industry grew, so did the participation of the Jatavs in it. Because shoe making was an occupational status traditionally assigned to the Jatavs, they did not suffer much competition from other castes, for whom such an occupation was precluded. In the early days of the industry Mohammedans also

A Jatav girl helps her father by making shoe heels.

entered into the industry to some extent as shoe makers, but more significantly as middlemen, or factors (*arhati*), in charge of market distribution. However, today the labor force of the shoe industry is by and large Jatav. One survey, made in 1960, with a sample of 1,351 workers in factories found that 85.8 per cent were Jatavs and 14.2 per cent were Muslims (Sisodia 1960: 19).

The sociological effects of these economic changes were many. At first the early contractors found that they had some leisure time as well as some economic independence and security. They were not bound to upper castes by personalistic patron-client (*jajmani*)[2] relationships (though it is possible that the contractors had such relationships with members of their own caste). Because of leisure, independence, and security, these wealthier Jatavs had the time to engage in activities other than bread winning. The advent of the shoe industry added a significant dimension to this situation. Since the making of shoes was at the outset almost completely in their hands, the Jatavs became occupationally as well as residentially segregated. Thus, interaction with other castes was reduced to a minimum. The observability of the Jatavs' role performance by other castes was lowered and consequently upper caste control over them loosened. This also meant that patterns of socialization could grow which were, in upper caste eyes, deviant or non-conforming to the traditional Camar place in the caste system.

A number of Jatavs were able to build and own their own homes, and some were even able to buy their own land. Members of the caste were, then, becoming an urban yeomanry. While there were within the caste two economic classes, that is, the minority of "big men" (*bare admi*) or wealthy Jatavs, and the majority of craftsmen (*karigars*) or poor workers, the position of both within the total society was low because of their common Untouchable

---

[2] The *jajmani* system is a cooperative system of production, distribution, and consumption. In it lower caste laborers and craftsmen worked for an upper caste patron and landowner. Membership in the system was hereditary and payment was in shares of grain, gifts of clothes, and so forth, which were distributed at set customary rates. In the traditional system, patron and clients were bound together not only by economic ties but also by ritual and political ones. For a recent, comprehensive, and analytical view of the *jajmani* system, see Kolenda (1963).

caste status. The situation remains much the same today. Moreover, wealth does not seem to remain for long in families. A
wealthy grandfather may live to see his grandsons impoverished.
Why is this so? One reason is that the division of inherited family
property among sons often reduces profitable, large, family-run
units into smaller, less viable units. Another is that sons of wealthy
men too often prefer to enjoy the fruits of their elders' industry
than to preserve them by their own efforts. On the one hand, this
continual circulation of wealth within the caste does not allow
for the permanent establishment of two classes within the caste
and, therefore, acts as a leveling process, leaving all with similar
life chances. On the other hand, it militates against the establishment of larger, and perhaps more capitally diversified, units which
might ultimately increase the economic power and standing of
the caste as a whole.

Until the time of partition, relations in the shoe market between the Jatavs, as producers, and the Muslims, as distributors,
seem to have gone quite smoothly. The Mohammedans, it seems,
ran few factories.[3] They remained sellers of raw materials (*malwala*), shoe makers, and factors. One informant said that possibly
this was due to their not wanting other men in or near their homes,
serving as factories, where their wives were in *purdah*. Be that as it
may, the Mohammedan factors were known to give cash for their
purchases and to advance loans on more favorable terms than are
available today. It is also said that they tried to get a better price
for the cottage shoe producer from the retailer. Furthermore, they
were supposed to have judged finished shoes according to their
worth and not strictly according to the profit they could make
from them.

In structural terms, then, organization of the shoe market was
divided between the Muslims who as factors performed the distributive function and the Jatavs who as factory owners (*karobar*)

_____
[3] The Census of 1911 lists 8,571 Hindus engaged as shoe and sandal makers,
while there were only 626 Muslims and 44 Christians in the same occupation (India.
Census Commission 1912: 589). I am at a loss to explain the discrepancy between
these figures and those given in footnote 1, page 32.

and craftsmen performed the productive function. The status-set of Muslim and factor was mutually exclusive of the status-set of Jatav and factory owner.

After independence, many Muslim factors emigrated to Pakistan; Hindu refugees from Pakistan, popularly known as Punjabis, came and took their place. Some of the reasons for the Punjabis gaining control of the market should be noted. A number of them had already been in the retail business and possessed both capital and mercantile acumen. With their capital they were able to buy at auction the stalls in the market vacated by Muslims and to set themselves up in a new business. With their entrepreneurial skills, they were able to make a success of their new occupation.

The Punjabis have bit by bit redefined the role of factor vis-à-vis the producer or Jatav. Today, instead of cash the Jatav frequently gets a credit slip payable after three months. The factor demands that, since he must wait three months for payment from the retailer, so too must the producer wait. Therefore, since most of the producers work on a hand-to-mouth basis, they are forced either to sell the credit slip at a discount to a money lender, or they must buy new materials at a higher rate because they purchase on credit. The Punjabis also attempt to get shoes at the cheapest possible price from the cottage workers, or producers. In order to make some profit, the producers then try to substitute substandard goods, thereby ultimately weakening Agra's reputation for shoes (Sharma 1958: 59).

Perhaps the most significant change in the whole structure of the market is that the Punjabis themselves are becoming factory owners. They are attempting to gain control not only of distribution but also of production. They thereby hope to break the Jatav monopoly in that area. This fact, in addition to the exploitation of the shoe craftsmen as producers, has generated much fear and resentment among the Jatavs, who understandably view it as a threat to their one source of security and economic advancement. In the hands of politicians, this fear becomes a potent issue. As one Jatav Member of the Uttar Pradesh Legislative Assembly put it: "If the Punjabis gain control of the factories, then it will be difficult to organize a united front against the factors in the market as we once did in the past when we went on strike against payment by the producers of the F.O.R. tax." This Member is assuming that there will be many scab workers if such a strike takes place. Considering the hand-to-mouth existence of many workers, it is a very real possibility.

The Punjabis' attempt to control the means of production by becoming factory owners is a cause of conflict between them and the Jatavs for two reasons. First, the status of shoe factory owner is traditionally ascribed to and occupied by Jatavs alone. Because the exclusive occupation of this status is economically advantageous to them, the Jatavs want to keep it that way and to retain their exclusive ownership of the means of production. If the Punjabis can make the status of factory owner exclusively theirs, the Jatavs will be reduced to the status of craftsmen, totally dependent upon factory owners no longer of their own caste. Even now, under factory owners of their own caste, craftsmen are at a disadvantage because there are no unions.

The second reason for conflict is that the Punjabis by becoming factory owners are gradually undermining what was formerly an exclusive Jatav occupation. Yet, there is no guarantee that the Jatavs can or will compensate for this by entering the profitable status of factor. This is due to the fact that Jatavs lack the business skills and capital to become factors. Since the shoe industry and market is not a highly modernized system of production and distribution, the business and entrepreneurial skills required are of a more traditional variety. They are learned more by socialization within a family and caste than by instruction in a school or institute. One wealthy Jatav did open a factor's stall in the market, but he failed because he lacked the necessary skills. Under such conditions, the Jatavs are at a distinct disadvantage, for the social interactions wherein these skills are learned are out of bounds to them. In such places their Untouchable status is salient and precludes structural observability over such upper caste private behavior.[4]

Furthermore, Jatavs will probably not be able to occupy the status of factor because this status is traditionally ascribed to and exclusively occupied by upper castes. Upper caste Punjabis resist any incursion on their occupational monopoly by the low caste Jatavs. The Punjabis' greater capital and control of the market system, including the retailers, make it easier to freeze out a lower caste competitor by making salient the Jatavs' Untouchable status within the market networks. Since shoe making and the shoe market are in the private sector of the Indian economy, there is in it no policy of "protective discrimination" as there is in the public sector, where the status of Scheduled Caste or citizen can be made dominant. Theoretically, if not practically, it would be difficult to prove discrimination because of caste in a competitive market system.

The situation was summed up in a letter written to me by a Jatav critical of his own caste:

Mostly Punjabis are educated and they know how to handle business. That is why they are rich and experienced owners in the business line. My community's people are not careful about their business. They spend their time in useless things, but among Punjabis this is not so. Another reason why Punjabis are prospering more than my caste is because Punjabis help each other in business and in other affairs. But my community's people are not so. . . . This may be true, that Punjabis may hold more factories in comparison to Jatavs because they are interested in business and they have money, which is the main thing for business, and they are educated also.

Distribution

The distributive centers of the Agra shoe market are in two places, Hing Ki Mandi, the larger, and Tikonia Bazar, the smaller, near Jama Masjid. Hing Ki Mandi has two market places for finished shoes, the Agra Shoe Market and the National Shoe

[4] See Goffman (1959: 112–13).

Market. The market at Tikonia Bazar is known as the Tikonia Shoe Market. Thus, there are three market places in all.

The main street of Hing Ki Mandi is lined with stalls where leather of all types and supplies, such as polish and tacks, can be bought. There are two other places just off Hing Ki Mandi Road where leather scraps, leather board, and canvas lining are for sale. One is at Golden Gate, where a spot on the ground may be had for thirty-six or fifty *naye paise*[5] per night. The other place is in the center of Hing Ki Mandi. It is a large open field surrounded on three sides by small stalls. Every evening in both places the ground is taken over by small dealers who, in a flea market atmosphere, sell leather scraps and pieces, while larger dealers occupy the stalls in Hing Ki Mandi. The leather bits and pieces of various sorts and sizes are used for heels, leather board inner soles, sandle straps, and inner linings of a shoe. Not a particle is wasted.

In addition to these markets there is the country (*deshi*) market, held behind the Agra bus terminal on Monday and Thursday mornings from about seven o'clock to twelve noon. Here, too, the same kinds of scraps and various other supplies are sold. People from the villages as well as city dwellers come here to buy, sell, and gossip.

There are four types of people who come to and interact in these market places. The first are the producers, who are either owners of shoe factories or craftsmen who work in them. Factory owners are of two types, contractors (*namewalas*) and basket men (*daliawalas*). Contractors are those who operate on a contract basis for the commission merchants or factors. When a factor has a retailer who places an order for shoes, he "puts out" the work to contractors. The contractor in turn organizes a group of workers to make the shoes. Basket men are those who produce shoes without an order. They bring their shoes to the market and auction them to the commission merchant who will pay the highest price.

---

[5] There are 100 *naye paise* in a *rupee*. A *rupee* was worth about 21 American cents during 1962–64 when this study was conducted.

Hing Ki Mandi, (above) main distributive center of the Agra shoe market. Men (below) having a smoke before business starts in the leather market of Hing Ki Mandi.

The second type of person in the market place is the commission merchant or factor. He is the middleman who buys from the producers and sells to the retailers.

The third type is the retailer (*bopari*). These are the men who come to Agra to buy shoes in the stalls of the commission merchants. And finally, there are the raw material suppliers (*malwala*), who supply raw materials to the producers.

The pivotal position of the factor in the market network becomes clear when his share of the total profits is considered. A reliable informant estimates the total cost and profits of a shoe as:

63.5%	materials for the shoe and cost of manufacturing
5.0%	profit to producer
6.5%	profit to factor
25.0%	profit to retailer
100.0%	

This estimate is fairly close to that of a survey (Sharma 1958: 91) taken in 1958, which reported costs and profits as:

66.67%	cost of shoe
3.33%	producer's profit
4.44%	railroad shipping charges
5.55%	factor's profit
20.00%	retailer's profit
99.99%	

If the producer distributes his own shoes, the profits and costs are:

69.21%	cost of shoe
6.58%	producer's profit
4.21%	railroad shipping charges
20.00%	retailer's profit
100.00%	

These figures seem at first glance misleading, since the retailer apparently gets about 20 per cent, the middleman about 5 per cent, and the producer about 5 per cent of the final sale price of the shoes. However, when one considers that a small contractor can make and sell only a dozen or so pairs of shoes in a week, while a factor can sell hundreds of pairs in the same time, little doubt remains as to who gets the lion's share.

The small profit of the producer is also whittled away by the heavy burden of taxes he must shoulder. First, there is the *katauti,* which is subtracted at the rate of approximately .015 to .020 *naye paise* per *rupee* from the amount paid the producer for his shoes. This tax is ostensibly for the services the factor renders to the producer or cottage worker. Actually, it is for the benefit of the factors, who put it in a general fund for charity or for other communal purposes. The producers have no share in this fund.

In addition to the *katauti,* the brunt of the state sales tax falls upon the small producer. He pays at the rate of .02 *naye paise* per *rupee* for raw materials. The factor avoids paying taxes by putting his tax on the retailer's bill. The retailer in turn bills his tax to his customers.

A final word must be said about the system of advances. Frequently, skilled workers are given salary advances which are, in effect, interest-free loans. The reason for this is twofold. First, when a man wants to entice a skilled worker into his factory, he pays him an advance, which the worker pays back when and if he can. Second, workers take advances because they are often in need of immediate cash for a marriage or for some illness in their homes, for example. The advance system, then, binds a man to the factory owner until he can repay the advance, which rarely occurs.[6] One factory owner told me he had put out about Rs. 14,000 in advances. Such a system not only keeps the worker in a type of debt servitude, but also ties up a great amount of capital, leaving it unavailable for more profitable investments.

Aside from its purely commercial functions, the market has other latent functions as well. It is an integrating institution

[6] I know of one case in which a man was in debt for 400 *rupees* for thirty-seven years. The money had been borrowed when he was a young man in order to perform his marriage.

through which producer, distributor, and retailer are enmeshed
in a national market network. The Jatavs know that their liveli-
hood depends upon leather that is cured and treated in Madras,
Calcutta, and Kanpur. Their wages are determined by the market
system and the forces of supply and demand on an all-India and
even an international scale. The market, then, is a leveler of
traditional statuses and sentiments to the extent that those who
interact within it do so as buyer and seller in a self-regulating
market of at least national scale.

Second, the market is a center of communication both within
the city, and between the city and hinterland villages. It is especially so for the Jatavs who meet friends there in a tea, sweets, or
wine shop, and often pass a few minutes gossiping on the street.
Third, the market is a center of socialization. Boys come there
carrying on their heads their fathers' shoes in a basket. They learn
what is involved in making a living in a national market network.
Fourth, the market emphasizes caste differences. The Jatavs are
factory owners or craftsmen and the Punjabis and Muslims are
factors and retailers. The economic advantages that attach to a
higher caste status are thus apparent and are increasingly resented
by the Jatavs. In post-independence India, equality of status and
opportunity as citizens is the overt creed, but inequality of status
and opportunity for members of unequal castes is often the covert
practice; in short, caste is still the dominant status.

Production

There are no comprehensive surveys on the number and types
of factories present in Agra City. To my knowledge, there are only
three large factories using machinery and capable of large-scale
daily production in the city.[7] These are all primarily engaged in
the production of police and military boots. Aside from the
mechanized units, most other units are of the small-scale industries
type, producing handmade shoes. These fall into two categories.

---

[7] This, therefore, excludes Dayal Bagh Industries, which are not included within
the city corporation limits and are thus not within the scope of this study.

The first are the organized factory units. Only forty-eight such units were registered with the government as small-scale units in the year 1962. However, they are estimated to be in the hundreds (India. Development Commissioner, Small Scale Industries, 1956: 32) and have some capital investment in tools, workspace, and materials. They usually employ fifteen or more workers. The group of Jatav leaders known as "big men" are generally the proprietors of such units.

Unorganized units compose the second category. These are found all over Agra City and its environs and generally employ less than fifteen men. They are located in the homes of work organizers or contractors, and have little or no capital investment in materials or workspace. These are owned by the contractors and basket men already mentioned.

In the larger factories the official employees on company books are really contractors who are assisted in whatever process they supervise by one or more "invisible" workers. This is a way to conceal, at least on paper, the actual number of employees and the use of child labor, both of which are subject to labor laws.

All work is on a piecework basis. It is not unusual for a worker or a contractor and his team to put in ten to twelve hours a day. However, as one large factory owner put it: "In a factory with machines, one can work only so many hours a day; here we can work longer hours because the pace is easier, the men stop for a chat, or for tea, and so on." The factories, then, are not only producers of shoes; they are also foci of the caste communication network within the city.

Among the unorganized units, Jatavs have a definite desire to be contractors. The reason for this is that one then has a more reliable source of income, and a better price for his shoes. The basket man must auction his shoes in the market for whatever price he can get, and because of his need for immediate cash the selling price is sometimes lower than the total cost of production. The contractor, on the other hand, contracts to manufacture shoes for a set price. If his work is good he will often establish a continuing

A typical morning at the country marketplace next to the Agra bus terminal.

Factors' shops
in Hing Ki Mandi.

Outside a small
unorganized shoe factory.

relationship with the factor, who will "put out" work to him when he has orders. Nor is initiative lacking here, for frequently both contractors and basket men will try to produce a new style of shoe and show it to the commission merchant or factor as a sample. If orders are placed on the sample shoe, the cottage worker profits from his initiative.

Shoes produced by the larger organized units are generally sold directly to retailers by agents or by one of the partners in the business. A number of these units supply shoes to large Indian companies, such as Bata and Flex. A few, usually family businesses, have opened retail stores in Agra, Delhi, and other cities. A crude estimate made in a survey undertaken by some Jatavs themselves states that there are about 3,000 fabricating units of all sizes in Agra, and of these only 150, or 5 per cent, have direct marketing practices (Bharatiya Juta Grih Utpadak Sangh n.d.: 5).

Shoe making in Agra is more than a caste occupation; it is a family affair and a way of life. Almost every male child grows up knowing at least the essentials of making shoes; it is only the rare educated or rich young man who has not, at one time or another, put his hand to the cobbler's thread and awl. Most children pick up some knowledge of the trade in their own or in their neighbor's home. However, when a boy leaves school in his early teens (the rule rather than the exception) he attaches himself to a man who becomes his *guru*, or teacher.[8] Nowadays a man's *guru* is most often a contractor who has his own shop or works in a larger factory owned by another man. The relation is symbiotic. The *guru* is helped in producing his piece work while the boy is taught some aspect of the trade.

Every year at the Festival of Rama's Victory (*Dushera*), new and old pupils are obliged to return and pay their respects to their

---

[8] I had my own adopted *guru* in a Jatav section of Agra City. He was the oldest man in the neighborhood (*mahalla*) and greatly respected. My position aroused no hostility; rather, my respect for him was interpreted as respect for the whole neighborhood. This relationship, plus that of an adopted sister (an old widow), were taken in good humor by all. Not only did it aid rapport, but it also enhanced my own feeling of participation in the life of the neighborhood.

guru. They give him some cloth, and always some betel leaf and nut (*pan*) and sugar candy (*batasa*). A man may have had more than one *guru*, but there is always one principal *guru* who gave him a start in shoe making and taught him his basic skills.

The learning process continues through a man's life and, if he is industrious, he never stops trying to pick up new skills. Nowadays some of the smaller cottage workers prefer to send their sons to the larger factories in the city, because there they can learn the latest techniques and innovations. The Russian exports require more skill than ever before, and not all men acquire the craftsmanship to make shoes of such high quality.

While the *guru* is a teacher, he is not always a master craftsman (*mistri*). A master craftsman is one who has mastered all the techniques of shoe making, including the design of new styles. Master craftsmen are at a premium in Agra, and there is always some competition among the larger factory owners for their services.

The standard wages paid for the various processes[9] of shoe making are given in Table I.

[9] The actual process of shoe making in Agra can be broken down into its component parts according to the names given to the different types of workers. They are:

Cutter. The cutter takes large pieces of leather and, using a pattern, cuts out the various shapes needed. Much time is spent estimating the maximum number of pieces in an uncut hide.

Fitter. A fitter sews the various pieces of leather cut from patterns to form the upper part of a shoe. A foot-powered sewing machine is used for this purpose. The work is considered to be one of the most skilled tasks and the pay is somewhat better.

Completer. This man will take the upper and put it on a last and then attach it to an inner sole. A mixture of finely powdered wood or cork is then put on the bottom of the shoe to make it level and smooth.

Soler. This man attaches the sole to the rest of the shoe by hand stitching with an awl.

Heeler. This man attaches the heel to the shoe. This must be done in such a way as to make the heel level. It is accomplished by building up with small pieces, if it is a cheaper shoe.

Taraser. This man smoothes out the sole and the edge between the upper part of the shoe and the sole.

Finisher. This man adds polish to the shoe and tries to conceal whatever defects might be present in it.

TABLE I. *Standard Wages for Skilled Labor*

Work	Wage per pair	Possible daily output	Possible daily wage
Cutter	Rs. .12	100 pair	Rs. 12.50*
Fitter	.36	1 dozen	4.50
Completer	1.50	4 pair	6.00
Soler	.12	3 dozen	4.50
Heeler	.12	3 dozen	4.50
Taraser	.18	3 dozen	6.75
Finisher	.36	1 dozen	6.50

* The cutters' wages, like all the others, are not always this much, since they are limited by the size of the order. If, thus, the order were for three dozen, his wage would only be 4.50 a day.

These figures are somewhat misleading, since real wages depend upon the season of the year, the skill of the worker, and the amount and kind of orders a worker or his factory owner gets. During the rainy season work is very slow, except in the largest factories. Furthermore, a worker frequently takes a day off, and both holidays and illness interfere with steady work. The best master craftsmen may make up to Rs. 250 a month, while the ordinary worker may make between Rs. 60 and Rs. 90 a month when he works steadily. The owner of one of the largest factories in Agra estimated that the average wage of his workers was about Rs. 2.75 to 3.00 a day. For a six-day week, this would be between Rs. 66.00 and Rs. 72.00 a month. This same factory owner allowed me to look at some of his salary payment books. The average monthly wages of ten men, who worked throughout the year[10] in his large organized factory, are compared below with those of ten workers in unorganized factories. The wages of those in the or-

[10] My original "respondent group" included every fifth man in the book. This resulted in a sample of forty-five workers. However, of the forty-five, only fifteen had a record that could be followed for a year. Of these fifteen, only ten were single men or a man and sons who could be used for comparison with the wage-earners of Bhim Nagar. The book again pointed up the problems of getting accurate wages over an extended period of time. Not only is work intermittent because of orders, but also workers may move from one factory to another, or they may take up cottage production at home, or they may be absent for sickness, festivals, or personal affairs.

ganized factories are on a high income scale which few men achieve. I was told that all these men were either solitary workers or men working with their sons, although I could not check the veracity of this information. These men were also highly skilled workers who were making shoes for Russian export. Their wages check with the possible daily wage scales presented in Table I. The average monthly wage for such a worker who works steadily over the year is Rs. 183.75.

TABLE II. *Average Monthly Wages*

| | Type of Factory* | |
Month	Organized	Unorganized
Jan.	Rs. 218.94	Rs. 82.03
Feb.	204.01	65.13
March	171.16	77.78
April	142.98	53.78
May	150.02	76.98
June	178.12	21.22
July	196.07	18.30
Aug.	170.75	25.10
Sept.	158.22	63.46
Oct.	224.74	76.82
Nov.	219.24	43.58
Dec.	168.76	63.58
Average Monthly Wage	183.75	55.64

* These figures are not strictly comparable, because those from the organized factory unit run from April of 1963 to March of 1964, while those from the unorganized factory units run from August of 1963 to July of 1964. However, they were the best available to me at the time. For a more detailed breakdown of these figures see Tables A-I and A-II of the Appendix.

The ten workers from small unorganized factory units were poorly skilled workers.[11] The average monthly wage for such workers is Rs. 55.64. Most such workers are subject to seasonal variations in employment. During the rainy season, from April to June, work

[11] My original respondent group was twenty-five men. However, because some men moved away, and others decided not to be available or just to "forget" the amount of their wages, the final respondent group was narrowed to ten. To follow up this group often took a whole Sunday morning's time.

dwindles to a trickle for some and to nothing for others. The peak season for work is during the winter, from September to March. The workers in unorganized factories are capable of earning much more than they actually do, but they are prevented from it by the seasonal variations in available work.

In terms of the minimal family budgets presented in Table III, it appears that either wages of independent workers are somewhat higher than I was told, or that these men are forced to borrow additional funds. It is also possible that there is an undisclosed source of additional income such as working wife or child.

Family I is composed of a man, his wife, and five children. He did not want to admit that his wife earns a few *naye paise* a day by selling fruit. Allowing her an estimated earning of Rs. 7.20 a month, his total income would be Rs. 67.20, if we accept his own estimate of his earnings as Rs. 60.00 a month. This leaves Rs. 11.20 unaccounted for. Since he is not in debt, he must be earning more than he admits. However, he is considered poor, and this budget was considered a minimal budget by other Jatavs. He is a laborer in another man's unorganized factory.

Family II is composed of two men living and working jointly, their two wives and nine children. Out of their earnings they claim to save enough to keep them through the slack work season. Their earnings are somewhat better since they are contractors in the market system, and for this reason their work is also somewhat steadier.

In addition to the problems inherent in having a small income, many Jatavs are also plagued by debt. In a respondent group of twenty-five men, the following responses regarding the problem of debt were elicited. Seventeen of the twenty-five were in debt for some amount of money, the average debt being Rs. 928. They gave the following reasons for their indebtedness:

Expenses for food, shelter, work	15
Marriage expenses	7
Sickness in family	6
Death ceremonies	1

TABLE III. *Sample Budgets* (*1964*)

| | Monthly (*30 days*) Expenses | |
Article	Family I	Family II
Flour	Rs. 37.50	Rs. 60.00
Vegetables		
Meat	9.00	30.00
Pulses		
Oil (cooking)	3.60	7.20
Spices	1.80	3.60
Salt	.15	.30
Kerosene	.90	.90
Fuel	5.40	15.00
Soap	1.00	1.00
Indian cigarettes and betel nut	3.60	18.00
Children's sweets, etc.	.24	37.50
Debts payments	0.00	0.00
Liquor	0.00	75.00
Clothes	8.30	10.00
	Rs. 78.45	Rs. 258.00

Five of these men had borrowed from relatives; two had borrowed some from a Jatav friend and some from a money lender; and one had borrowed some from relatives and some from a money lender. The others had borrowed from money lenders at rates of 15 to 38 per cent per annum. Few people who go into debt for large amounts to money lenders are ever able to pay off the principal. Borrowing from relatives and affines is preferred, as it is usually done without interest and is easier to repay. Most men blame their indebtedness on the lack of steady work. The same respondent group of twenty-five men provided the following answers to the question, "In your opinion, what is the solution to the problem of debt?" (See Table IV, p. 53.)

The list in Table IV is instructive because of the variety of solutions it gives to Jatav economic problems. The need for steady work is no doubt the most important and crucial. But it is quite interesting to note the high proportion of the group who consider self-help through the elimination of "bad habits" such as drinking

TABLE IV. *Solutions to Indebtedness*

A. Steady Work		12
B. Self-help and hard work, while refraining from bad habits such as drinking and gambling		8
C. Government help by:		
1. government orders for shoes	1	
2. declaring equal wages for all	1	
3. remission of all debts	1	
4. advancement of capital	1	
5. control of moneylenders	1	
		5
Total		25

and gambling a partial solution to their problems. While it is true that Jatav men gamble and drink, as far as I could judge, it did not seem to be a great problem in itself. The real problem lies in the fact that liquor costs money, as does gambling, and to spend money for it under the present conditions is a luxury that few Jatavs can really afford. Their self-criticism, of course, also reflects the cultural value in India that drinking is degenerate, low caste behavior. It also reflects internalization of the upper caste stereotype that Jatavs are by nature drunken and misbehaved. A private report written by an upper caste government officer concerned with the shoe industry in Agra describes Jatavs in the following terms:

The workers are addicted to a number of vices. Gambling and alcohol claim a major part of earnings. They are constantly under debt which is never cleared. They remain absent from work, are often found dead drunk on weekends, and on Mondays they are often hunted out from their dens by the manufacturers. The result is that in spite of their good earnings . . . they remain miserably poor. . . . They usually live in slums with the result that their domestic life is not quite healthy.

When such views exist even among educated government officials, who are supposed to be in positions demanding at least some empathy toward the shoe makers, then the success of their development programs seems hazardous for reasons other than the alleged "habits" of the workers.

The third set of reasons (section C. of Table IV) indicates the variety of methods that the government can conceivably use to solve the Jatavs' economic problems. The importance of these reasons is that the Jatavs feel that the government *ought to do* these things; they are expected role behavior of the government. I was constantly asked, in essence, "Why doesn't the government help us?" This attitude toward the government may have its origins in the patron-client (*jajmani*) system of the villages, where the Jatavs, as Camars, were dependent servants of upper caste patrons. It is not inconceivable that rural patron-client attitudes have now been displaced on the biggest patron of all, the government, whose Five Year Plans and socialistic goals have led the Jatavs to expect much. However unrealistic their expectations may be, Jatavs feel that the government has the power to effect a solution to their economic problems but does not do so, because it is controlled by upper caste people. They believe that their Untouchable caste status is made salient against them in dealings with the governmental representatives. Politicians who can convince the Jatavs that if elected they can solve their economic problems have here another potent vote-getting issue.

Many Jatav workers feel there is a need for a strong union to fight for their rights, but at present no such union exists in Agra. In 1937 an organization called the Shoe Workers' Union was established, but it failed to take root. Again in 1945, a Shoe Workers' (*Mazdoor*) Union was founded. It tried to bargain for the remission of advances made to workers during World War II, but the attempt failed. It, too, died out during the throes of Partition and its aftermath when some of the union's leaders fled to Pakistan and elsewhere.

There has been some talk in the Republican Party, to which most Jatavs belong (see Chapter IV), of organizing such a union, but the few half-hearted attempts to do so smothered in the quicksands of rhetoric. Horace, when speaking of would-be poets, aptly characterizes the situation: *"Parturient montes, nascetur ridiculus mus."* (They labor as though giving birth to giants [mountains], and bring forth naught but a squeaking little mouse.)[12]

The Role of Government

Government help to the shoe industry and the shoe makers has little profited the small producers. In 1955, a report on Leather Footwear in the Northern Region of India by Dr. Eugene Staley, Ford Foundation Consultant, and S. Najundan, Deputy Development Officer, was submitted to the Development Commissioner and Director of the Northern Small Scale Industries Institute. This report, among other specific recommendations, stated that:

The first step is the training of a group of footwear industry extension officers. These officers must have a good knowledge of the techniques of footwear manufacture. . . . They must also have the type of personality and the necessary training to make them good extension workers (India. Development Commissioner, Small Scale Industries, 1956: 5).

While this particular recommendation has not been carried out, there is now a branch of the Small Scale Industries Institute at Agra. The Institute is developing a program with three specific goals: to train workers who already know all the processes involved in shoe making but lack a scientific and efficient outlook; to develop managers who will benefit from modern management practices by utilizing the schemes and plans which the government develops for their use; and to carry out industrial surveys of trends and prospects. The benefits of this program have not yet trickled down to the small-scale producers as a whole. My impression is that these programs are becoming an end in themselves, rather than a means to an end.

One young man who had interited Rs. 12,000 from a distant grandmother sought help from an official of this institute in managing and investing his inheritance. He told me:

[12] Total account given in the last two paragraphs is substantially from Upadhyay (1951: 81–82).

The reason I have succeeded is because of the help of Mr. X and because I'll take a risk. At first, I had an ordinary shoe factory of my
father's. But then I took a chance on Russian orders and succeeded.
X is my *guru,* and he says I am the only one he has ever really gone
out of his way to help. He at first tested me for three months by giving
me appointments for which he was never present. When he saw I was
determined, then he gave me all the benefits of his knowledge, experience, and contacts.

While one cannot generalize from this statement, it suggests the
following observations. First, communication between government
officials and factory owners is neither very great nor very much encouraged. Second, "help" might be as much in terms of contacts
which can be made as it is in terms of knowledge and experience;
for this reason it is not liberally dispensed. Third, entrepreneurship
in terms of taking risks is present and can succeed, especially when,
like this young man, the entrepreneur is both literate and enter-
prising. In view of the Staley report, such entrepreneurial types
possess an adaptive advantage in an increasingly competitive market. The report notes:

The main hope for expanding the market and raising the level of earnings of cottage workers lies in technological improvements. Only thus
can the existing industry become competitive in quality as well as
price, and only then can it maintain its position over a long period and
capture new markets, especially export markets (India. Development
Commissioner, Small Scale Industries 1956: 38–39).

Nonetheless, there is a general fear among the workers and an
ambivalence among the factory owners towards the introduction
of machine technology. One of the largest factory owners told me
he did not want machines "because machines are too expensive.
Even if the government helps to get them, it puts many conditions
on their use. They are too expensive to keep up." Yet, at another
time he indicated that machines would inevitably come. Thus, he
was investigating their possibilities. He said, "In ten years machines will come to Agra. Then, so, Jatavs will be out of work."
This was echoed by a perceptive worker, who in one breath con-

cluded, "Then we will all die" and in a second breath asked, "Can you tell us some other business to get into?"

The need for diversification into new industries was strongly emphasized in the Staley report. But few Jatavs are perceptive enough to take note of it or daring enough to look for alternatives. Among the college students there is a feeling that the shoe industry would be profitable even for them, if they had the capital or an inherited family factory with which to start a business. This attitude is buttressed by the fact that government service is the only other opportunity structure open to them.[13] In private industry the salient status of Untouchable continues to operate. One Jatav with a Master's Degree in Social Work told me that although there were many jobs in private industry for a man of his qualifications, he could not get one because of his caste. He was, therefore, undecided whether or not to go into his family's shoe business or into government service. Among the uneducated and the workers there is a feeling that in any other business they will fail; therefore, why try? "Who will buy sweets or tea or anything else from the [polluting] hand of a Camar?" is a constantly heard complaint, although in a city such as Agra not entirely justified.

In 1955, the government of the state of Uttar Pradesh started a pilot project with the purpose of bringing to the cottage worker "the advantages of the mechanical production in small units, to train him on such machines, and to do him service by improving his production" (Manager, Government Pilot Project, n.d.). A model factory with modern machines was established for this end. The project has, however, had little effect on the cottage industry worker, who has neither money nor need for machine technology, and it has done little to communicate knowledge to him of the

[13] In a note entitled "Backward Classes," the following appears in the *Economic Weekly* of Bombay: "Only 40 per cent of the reserved vacancies in jobs listed by Employment Exchanges were filled. In the Government, out of 17.75 per cent seats reserved for them as many as 16 per cent were filled, but the sad truth is that 78.6 per cent of the vacancies filled were for sweepers and 20 per cent were class IV jobs and the share of the backward classes in class I jobs was a mere one per cent" (*Economic Weekly* April 10, 1965: 622).

other facilities it has to offer. After the Chinese incursion into India, the model factory and its staff were switched to the manufacture of boots for the army. Thus, the Pilot Project is now part of a war effort, but not the war against Jatav poverty and ignorance as was originally intended.

Another scheme initiated by the government is the Quality Mark Footwear Manufacturers' Cooperative Association, Ltd., begun in 1954. Its purposes were to get raw materials, to stimulate orders for shoes, and to prepare shoes of a standard make so that they could bear a "Quality Mark" of certification from the Uttar Pradesh government. The Association also was supposed to make arrangements for marketing and selling of shoes and to get raw materials at wholesale rates for its members. Today it is composed of thirty-two primary societies of shoe makers under one central society and is located in a large building on the Civil Lines near Agra College. The Association has a number of machines for making boots for the government services and has been assisted by government loans and grants. But only a few of the primary members profit from the enterprise. The president is a Brahman who is said to have achieved his position with the clandestine help of a Jatav president, who handed the presidency over to him. Therefore, the feeling expressed among some Jatavs is that "the property is Jatav but the rule is Brahman." The Cooperative thus seems to have benefited only the few who were able to get control of it. However, the Quality Marking scheme, which is separate from the Cooperative, has helped those who were able to become members and obtain a Quality Mark registration number from the government. Government advertising has, to some extent, raised the standard of shoe production in Agra and its reputation outside the city.

The National Small Scale Industries Corporation of the State Trading Corporation of India has attempted to expand the market of shoes through its export and domestic Cells. While the domestic cell has had much trouble, the foreign branch has successfully channeled orders from Russia to small-scale manufacturers in

One of the few remaining shoe repairmen who sits in front of the Agra bus terminal.

Agra, despite the many complaints about its system of inspection
and its delay in returning security bonds put up by the manufacturers against defective goods. The enterprise as a whole has benefited the workers to some extent by creating an export market for
Agra shoes. But again, work from this source is not regular and,
therefore, does not give steady employment. Nonetheless, it is this
organization which has integrated Jatavs into a national governmental structure which also operates on an international scale.

In October 1963, a meeting was held in Agra under the auspices
of the Indian Shoe Production Association (Bharatiya Juta Utpadak Sangh), an organization sponsored by a small group of Jatav
"big men" who support the Congress Party. These men used their
status as Congress Party Members, as well as their status of (organized) factory owner to get recognition from the government. Sit-
ting in at the meeting were the Deputy Registrar of Cooperative
Societies and representatives of the Cooperative Bank, the Small
Scale Industries Board, and the State Bank of India. The meeting's
purpose was to discuss the economic problems of small-scale shoe
manufacturers in Agra and to seek possible solutions to them. It
was agreed that the reason all previous cooperative societies had
failed was that they had been credit societies to which members
had never really paid back loans. An alternative solution seemed
to lie in the formation of a production-with-distribution unit. Such
a unit would put market distribution in the hands of the producers
and remove it from the hands of the Punjabi factors. It would also
provide for direct marketing to retailers. The production unit
would attempt to keep quality controls and apply new techniques.
This proposal was favorably received by the government, and
funds to start it were provided. Whether or not it will succeed
where other programs have failed is yet to be seen.

This meeting was instructive in another sense. The Jatav "big
men" who were present had some familiarity with the Staley report through the help of a Buddhist monk literate in English.
Some of their thinking was influenced by this report. They, then,
were an interest group lobbying with the government to enact the

recommendations made in its own report. The meeting was also an implicit bid for political power. By trying to bring home benefits for smaller factory owners, the Jatav Congress "big men" were also trying to bring in votes for Congress. No doubt, too, they were also trying to gain some prestige for themselves and their views.

This meeting is an illustration of the changes brought about by the introduction of a democratic system. Previously, these men would have received little or no attention because of their dominant Untouchable status; now, because of their dominant citizenship status, they can exert pressure as an interest group. These men have added to their status-sets within a particular caste the statuses of wealthy man, Congressman, and "big man." It is this sub-set of newly dichotomized[14] statuses that opens to them channels of communication and influence within the administrative and governmental structures. As Congressmen, they are members of the party now in control and are entitled to the patronage which goes with its support. The Congress government gives them patronage and a position in the public sector of the economy, while the rich men in turn give financial and voting support to the Congress. They also aid the administrators involved in government projects to submit "successful" project reports. Such reports naturally help the administrators justify their positions and better their chances for advancement.

In another sense the caste status of these "big men" has become "politicized" and has transformed them into a bloc of *voters* for the Congress Party, small though the bloc is. They offer the Congress a potential wedge into the predominantly non-Congress Jatav caste of Agra. This is important in view of the fact that the Jatavs are about one-sixth of the city's population. The policy of divide and rule evidently did not leave India with the British.

In addition to the cooperative societies, various other government agencies make available loans and grants, but the leather

---

[14] The specialized meaning of this concept has been explained on pages 16–19.

industry has been unable to avail itself of them. Loans given under the State Aid to Industries Act are "not given according to the requirements of [financial need of] the applicant" (Sharma 1958: 109). The State Bank of India started a pilot project to finance small industries through various agencies such as the State Finance Corporation, the Cooperative Bank, and the Commercial Banks. It finances the purchase of raw materials and other needs if the applicant can himself give a down payment of 10 to 15 per cent of the total cost. The State Bank has also promised credit to units receiving orders from the National Small Industries Corporation (Sharma 1958: 106–08). However, only the larger factories and the Congress "big men" can avail themselves of such loans. These "big men" with their Congress Party connections and activities are in a better position to obtain available governmental loans and grants. The smaller factory owners, because they have little or no credit rating with official agencies, find it virtually impossible to get such loans and grants.

Summary

The dominant status of Camar was adaptive for the Jatavs in the economic field until the post-independence years. The Jatav status of shoe maker (Camar), whether as factory owner or as craftsman, tended to remain mutually exclusive of the Muslim status of factor. As a result, the Jatavs were virtually in complete control of the means of production. This fact resulted in occupational isolation and independence from other castes. In the post-independence era, however, the Punjabis, who have taken the place of the Muslims, are attempting to occupy the status of factory owner, thus making it part of their status-set. When this occurs, the status of shoe factory owner is cross-cut by that of both Jatav and Punjabi. Consequently, Jatav economic independence and control of the means of production are threatened. Yet, there is little attempt on the part of the Jatavs to occupy the Punjabi status of factor. If the incursion of the Punjabis continues, most Jatavs will be reduced to the status of craftsmen and to economic de-

pendence upon other castes. Moreover, the opportunity structure
of the Indian society is for the Jatavs severely limited to the public
sector of the economy and to government jobs, since in the private
sector and in small business their Untouchable status can be made
salient as one more negative factor against them in a competitive
market.

The conflict between the Jatav producers and the Punjabi factors now has a double edge against the factors. The Jatav factory
owners react to the threat of the Punjabi factors' entrance into
their monopoly over the means of production, while the Jatav
craftsmen react to the Punjabi factory owners' attempts to exploit
them. The Punjabis, then, have become for the Jatavs an inimical
group, a negative reference group, whose power and influence
must be curbed. This has given rise to a serious political issue, and
the Jatavs see political action as a means to its solution.

Various government agencies are integrating members of the
Jatav caste into institutions at the urban, state, national, and even
international level. In these institutions the Jatavs occupy new statuses—such as, producer in the market; member of a cooperative in
government-sponsored activities; petitioner for a loan or grant or
advice on orders for Russia from the banks or in a Small Scale Industries Corporation—which have progressively dichotomized from
the status of Camar. In these statuses Jatavs interact with officials
in the government in a coordinate way, because they also have the
all-important status of citizen; this is in spite of the fact that these
officials might have a high caste status which in the caste system
would require a superordinate-subordinate relationship. These interactions actualize for the co-actors their position in a state and
nation which has legitimized and required by constitutional fiat
equality of status, or citizenship, for all. This is not to say, however, that informally their Untouchable status is not made salient
against them.

The greatest economic problem of the Jatavs in Agra today is
that of finding year-round work. With their skills they can earn
at least a minimal living wage. Yet, they are unable to control sea-

sonal factors, new competition from other areas of India, and the vagaries of supply and demand. The need now is for further rationalization and planning in the market system as well as for a more equitable distribution of profits. The unrest created by these conditions is another important source of Jatav discontent and political demands; this topic is treated in the next chapter.

An evaluation of the Agra shoe market would not be complete if it failed to point out some of the traditional cultural elements carried over into it. In the first place, the craft of shoe making is still embedded within the Jatav caste; that is, the status of shoe maker (Camar) is ascribed to the Jatav caste and is not yet open to (or even desired by) other castes. The art of shoe making is still passed down within families or by a Jatav *guru*. Learning of the craft from a *guru* and worship of tools at Dushera surround the craft with a traditional sacredness which resists profanation. Jatavs identify, and are so much identified, with the craft of shoe making that for most of them it is a way or style of life rather than a means to a way of life. The relationship of the factory owners to the craftsmen in the system of advances indicates how a simple and universalistic relationship tends to become complex and personalistic. Such a relationship ties up capital that might be invested in more profitable outlets.

In the second place, just as the craft of shoe making is embedded in the Jatav caste, so are the business skills of factors embedded in the castes and families of the Punjabis and the few remaining Muslims. Thus, the functions of production and distribution resist further rationalization and subjection to the criteria of efficiency and merit.

I have also noted that the Jatavs look to the government for solutions to many of their problems. While this can be interpreted to some extent as an expectation of the Jatavs for adequate wages and government assistance in a society which seeks, and to some extent is achieving, a socialistic pattern, it is not a completely satisfying interpretation. There is, in my opinion, the added factor of a displacement of traditional Jatav dependency upon upper castes to dependency upon the government. While this dependent behavior leads the Jatavs to accept the government's socialistic goals easily, it does not necessarily encourage initiative, independence, enterprise, and self-help among the Jatavs themselves. Thus, their definition of the government's socialistic role is broader than the government's own definition of that role. Since the Indian government operates in an "economy of scarcity" (Weiner 1963) and cannot possibly fulfill all Jatav expectations, these conflicting definitions become two more sticks to pound on the politicians' kettledrum.

On the whole, though, the present state of affairs is a far cry from the pre-1900 days, when most Jatavs were little more than laborers and city servants. The market and, later, the introduction of the institutions of democratic socialism have helped to expand their status-sets and have given them new means to relate themselves to other castes and individuals in India. Not all change has been, by any means, for the worse.

CHAPTER IV

The Politics of Untouchability

In the preceding chapter, it was suggested that in the hands of the politicians the economic problem of the Jatavs has become a part of the "politics of scarcity."[1] The economic problem is only one among many problems which underlie Jatav policies, demands, and affiliations in the political arena. Thus, for the Jatavs, the "politics of scarcity" is subsumed under "politics of Untouchability,"[2] a phrase which itself gives some clue to the contradiction that underlies it. Essentially it is the confrontation of groups tradition- ally defined and accepted as powerless pariahs with groups tradi- tionally defined and accepted as powered gentry. This confronta- tion is occurring in a new field of interaction where individuals, not groups, are defined as equal and have equal rights to the sources of power. The history of Jatav "politics of Untouchability" falls into three periods: The pre-independence period, the transi- tional period, and the post-independence period.

---

[1] Weiner speaks of the politics of scarcity in the following way: "The gap between government plans and decisions and the many demands of organized groups is so great that the danger is ever present that neither democratic institu- tions nor effective policy will survive" (Weiner 1963: xiv).

[2] Aspects of this have been treated in Lynch (1968).

The Pre-Independence Period

In the last chapter a picture of the changing economic status of members of the Jatav caste was drawn. The change, at first favorable, began at the turn of this century and continues even today, although recent changes appear to be more disadvantageous than advantageous. From the beginning, economic change for the Jatavs resulted in residential and occupational segregation, social isolation, economic independence, and leisure time to engage in other activities. Economic change also brought about a similarity of status attributes (wealth and the wealthy way of life), which provided a basis for reference group behavior by Jatavs, vis-à-vis upper and wealthier castes. "Once this minimal similarity obtains, other similarities and differences pertinent to the situation will provide the context for shaping evaluations" of self in front of others (Merton 1957: 242–43). This "assumes that individuals comparing their own lot with that of others have some knowledge of the situation in which these others find themselves" (Merton 1957: 247). It was precisely this knowledge of similarities and differences pertinent to the situation which the early Jatav contractors and later the Jatav shoe makers did not have, although they were wealthy. Entrance into the local Indian opportunity structure was closed to them because of their status as Untouchable; they did not have a ticket which would allow them entrance into restaurants, schools, temples, and so forth. Untouchable status denied them structural observability, and their knowledge of the situations in which other castes found themselves and played their parts was therefore much impeded.[3]

However, a significant change came with the growth of the mission schools and more importantly, with the growth of the Arya Samaj.[4] The Arya Samaj, whose leading members, according to

---

[3] "Since, in the traditional system, only the Brahmin priest was the repository of knowledge of the Great Tradition, the dominant caste was able to prevent cultural trespass by ensuring that the priest served only the high castes" (Srinivas 1966: 17).

[4] The Arya Samaj was a Hindu Reformist Sect begun by Swami Dayanand Saraswati in 1895. Its teachings included a distinctive interpretation of the most ancient of Hindu scriptures, the Vedas. It was against idolatry, the non-marriage

Jatav reckoning, were Brahmans and persons of other upper castes, not only taught the three R's to the Jatavs, but also imparted to them a knowledge of Sanskritic symbols, rites, and beliefs. Response to these new opportunities had begun before 1900 when the Jatav rich men Seths Sita Ram and Man Singh, came under the influence of a Brahman named Swami Atma Ram.

Swami Atma Ram had written around 1887 a book called *The Ocean of Knowledge (Gyan Samudra)*. In this book it was written that:

Among the gods, Indra, Shiva, Varun, Yama, Surya, Agni, Ashvani, etc. these ten are counted as belonging to the Kshatriya varna. If according to the Lomash and other Ramayanas the origin of the Jatav race [*vansh*] is traced from the gotra of Shiva, then how can there be any doubt that it [Jatav race] is from among the Kshatriyas? (quoted in Yaadvendu 1942: 98).

The Lomash Ramayana, it was said, could be found only in Nepal, whence came the Swami. Thus, the necessary, although not sufficient conditions, as we shall see, for the Sanskritization of the Jatavs were present at that time. There was economic independence and a modicum of wealth, leisure time to learn and propagate more orthodox behavior, and a scriptural discovery of a new identity. Suitably equipped, then, the Jatavs set foot to the path of Sanskritization. With Swami Atma Ram's help, they began to assert a claim to Kshatriya status. Moreover, the two Seths organized a Jatav Committee of Agra City which tried to bring about social reform among their caste brothers. A caste council (*panchayat*) of Jatavs from all of Agra City was convened, and it resolved to forbid the eating of beef and buffalo by Jatavs from that time forward.

During the first two decades of the 1900s, a number of the early

---

of widows, caste, and polygamy. Because of its reformed views and innovations, it found little favor with traditionalists, especially orthodox Brahmans. Part of its active goal was to bring back to Hinduism those who had been converted to Islam and Christianity. (See Griswold 1913: 57–62; Rai 1915; and for the Arya Samaj in Agra, Nevill 1921: 72.)

contractors' sons received some education in the mission, Arya Samaj, and government schools. In 1917, they banded together and formed the Jatav Men's Association (Jatav Vir Mahasabha). Later (1924) the Jatav Propaganda Circle (Jatav Pracharak Mandal) was formed.

These societies were intended to motivate the Jatavs towards education, Sanskritization of their way of life, and change of their identity. Education was considered of primary importance. Without it, the knowledge pertinent to effective reference group behavior and the behavior proper to higher caste rank would remain hidden behind the curtain of ignorance.[5] The first problem was to instill the Jatavs with an awareness and a desire to accept this new definition of their caste rank, and the means of legitimizing it. A member of the Jatav Men's Association complained that: "There are some Jatavs also who, being crushed by them [the upper castes], think it to be a great sin to educate their children. They do not understand that it is learning and education which makes mankind above all creation" (Sagar 1924).

In 1924, a book *Jatav Life (Jatav Jivan)* was written to justify the claim to Kshatriya status. This was followed in 1942 by a more sophisticated book, *History of the Yadu[6] Race (Yaduvansh ka Itihas)*, which was written by a Jatav lawyer. The writer of the earlier book made explicit the Jatav denial of their Untouchable status. He wrote: "Through this book we want to tell other castes that the Jatav race is one of the sacred and highest races and is not untouchable" (Sagar 1924).

What, then, is the sociological definition of this movement for upward mobility? To begin with, for the Jatav, there was more than one reference group. The Arya Samaj formed a reference group of imitation, since it was the Vedic[7] practices of this group which the Jatavs tried to imitate in behavior, rite, and belief. This

---

[5] Cohn (1954: 127) finds a similar belief among the Jaiswar Camars in Senapur village.

[6] Yadu is the same as Jatav, at least according to the Jatavs.

[7] Vedic practices are those which are supposed to be contained in the Rig Veda, the most ancient and most sacred of Indian scriptures.

meant that meat-eating and scavenging were to be prohibited and the practice of the Vedic rituals or counsels (*sanskars*), especially the sacred thread ceremony, were to be adopted. It is noteworthy that the Arya Samaj taught not only the Vedic way of life but also the belief that an individual's caste status was achieved, not ascribed. This teaching was taken by the Jatavs as a mandate for group action to change the status of the whole caste. The conception of the legitimacy of individual achievement was thus planted in the Jatav mind; with it came an awareness of an alternative form of social structure and social mobility. The manifest function of the Arya Samaj was to reconvert or bring back the Scheduled Castes to the fold of Hinduism. It did this by a process of religious enculturation which included an anticipated rise in caste rank. An unanticipated consequence of the Arya Samaj's teachings was a watering down of the caste ideology by those traditionally associated with its immutability, the Brahmans, although they were Arya Samajists. In other words, although the idea of legitimate individual achievement was not useful to the Jatavs operating in the caste system at that time, it was, nevertheless, useful to them as anticipatory socialization for the social system which was to come after independence.

However, the Jatavs did not identify with the Arya Samajists. Their reference group of identification was the Kshatriya *varna*. The *varnas* are the traditional four classes of Indian society: Brahman (priest), Kshatriya (warrior), Vaisya (merchant), and Shudra (menial). *Varnas* are not really social groups; they are merely social categories.[8] The operative units of Indian society are not the *varnas* but the *jatis* or organized and generally endogamous groups popularly known as castes, which number in the thousands.

There is here an important point to be grasped. Some students

---

[8] I use the term social category to mean a classification or label for a type of people. Such a group has neither organization, nor common purposes, nor common sense of identity which characterize social groups. For example, juvenile delinquents are a social category, while the Black Zephyrs of Slumtown are a self-conscious and organized group among the many groups or gangs labeled delinquent.

of Indian society argue that the *varnas* are meaningless or, at most, are very general all-Indian categories into which castes can be classified. It is the *jatis,* they say, which are the only important social groups or categories in contemporary India. Yet it should be clearly evident that the *varnas* are very important as social categories which allow for some elasticity in what otherwise appears as a rigid hierarchy of castes. The *varnas* are open categories in which the principles of recruitment are both ascriptive and achieved. It is precisely for this reason that socially mobile Indian castes have used and still continue to use them as reference groups of identification. A mobile caste does not claim membership in a *jati,* since the rules of caste endogamy prevent this. But it does claim membership in a *varna,* because there is no question of caste intermarriage or assimilation involved. Even as far back as 1885, Nesfield noted that: "In Upper India the manufacturing of Chattris is a process still going on before our eyes, and what is happening now has been in operation for the last two thousand years at least" (Nesfield 1885: 18).

The aim of Jatav self-identification as Kshatriyas was not assimilation and intermarriage, rather it was a change of dominant status and therefore of rank within the caste system. The crucial task was to prove an acceptable genealogy which would legitimize their claim. It was asserted that the Jatavs were of Yadu Race (*vansh*) and thus were really of Kshatriya *varna.* This was explained by stating that Jatav was merely a dialectical variant of Yadav; and Yadav was a modern variant of Yadu. Since the Yadus were an ancient Kshatriya tribe, and the Jatavs were their modern descendants, they were, therefore, of the Kshatriya *varna.* Further proof of Kshatriya ancestry was found in a series of correspondences, or status similarities, between the Jatavs and the Yadav (Yadu) race of ancient India. These included identical *gotras*[9] and such Kshatriya-like ceremonies as shooting a cannon at weddings and the use of the bow and arrow at the birth ritual (*san-*

[9] *Gotras* are named, exogamous groupings of kin.

skar). These similarities of status attributes were the contempo-
raneous basis on which the identity of Jatav and Kshatriya status
was claimed and taken as proven, at least to the Jatavs.

However, once the claim was made that they were a separate *jati*
in the Kshatriya *varna,* the Jatavs had to give some explanation of
how they had become Camars. For this purpose the legend of the
Brahman Parasuram[10] was pressed into service; it was interpreted
to mean that Jatavs were really survivors of the ancient war between this Brahman and the Kshatriyas. The occupations of shoe
maker and contractor were rationalized as a defense and a disguise
to which Jatavs were reduced in order to escape the wrath of
Parasuram. "The main reason for the non-availability of our complete history is the jealousy of Parasuram and his disciples towards
the Kshatriya [warrior] race" (Sagar 1924).

In this way the Jatavs tried to solve the two identity problems
faced by a mobile group in India (and in my opinion by a mobile
group or individual anywhere). First, there was the problem of
legitimation or proof of their new identity. They attempted to
solve this problem by appealing to the authority of scripture in the
Lomash Ramayana, of myth in the Yadav genealogy, and of custom
in the bow and arrow ceremony, and so forth. Second, there was
the problem of explanation, or proof of how their present identity
was a mistaken one; this they attempted to solve by appeal to the
Parasuram myth. Srinivas[11] has implicitly recognized these two
problems and has, moreover, noted a third. This is the problem of

---

[10] It is interesting to note that "the city of Agra is supposed to have been the
birthplace of the *Avatar,* or incarnation of Vishnu, under the name of Parasu
Rama, whose conquests extended to and included Ceylon" (Hamilton 1828: 18).
Parasuram was a mythical Brahman who appears in the Mahabharata and the
Ramayana. His hostility to Kshatriyas was so great that he is alleged to have
cleaned the earth of them twenty-one times (Dowson 1961: 230–31).

[11] Srinivas has written: ". . . Sanskritization does not automatically result in the
achievement of a higher status for the group. The group concerned must clearly
put forward a claim to belong to a particular *varna,* Vaishya, Kshatriya, or Brahman. They must alter their customs, diet, and way of life suitably, and if there are
any inconsistencies in their claim, they must try to 'explain' them by inventing an
appropriate myth. In addition, the group must be content to wait an indefinite
period, and during this period, it must maintain a continuous pressure regarding
its claims" (Srinivas 1962: 57).

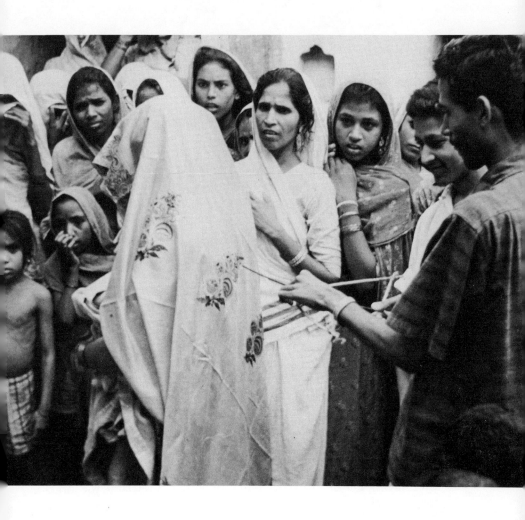

Presentation of bow and arrow at birth ceremony, a symbolic act upon which Jatavs claimed Kshatriya status (see page 71).

perseverance or iteration and reiteration over time of one's claims until accepted. This the Jatavs attempted to solve through their organizations such as the Jatav Men's Association.

The final element in the structural definition of the Jatav attempt at social mobility was pinpointing the negative reference group. This was the orthodox (*sanatani*) Brahmans who were the enemy in fiction as the descendants of Parasuram and in reality as the rejectors of the Jatav claim to Kshatriyahood. While other castes also rejected the Jatav claim, it was the Brahmans who, in Jatav eyes, were really responsible for the behavior of these other castes toward the Jatavs. They felt that the Brahmans as priests both wrote and enforced the rules of caste morality which were found in the Hindu scriptures and which made them Untouchable.

It was, then, these three reference groups of identification, of imitation, and of negation which defined the social situation for the Jatavs. They provided, as it were, a map of social space as the Jatavs saw it; yet, it was a map whose contours and boundaries were drawn differently from those of other castes. Thus, conflict arose over which version was correct.

The conflict in this situation was between the Jatavs' definition of themselves as Kshatriyas and the others' definition of the Jatavs as Untouchables. These others were higher in the caste hierarchy and rejected the Jatav claim to a similar rank by activating the Jatavs' salient status of Untouchable. In structural terms the Jatavs were laying claim to a new status of Kshatriya which the other castes would not accord. Were the claim to Kshatriya status accorded by other castes, two things would structurally result.

First, all those statuses which were open to Kshatriyas, whether ascribed by birth or obtainable by achievement, would be open to the Jatavs without mobility-induced conflict. If Jatavs were accorded the controlling status of Kshatriya and thus the legitimate right to occupy other statuses in the status-set of upper caste Kshatriyas, then there would be no cause for conflict. In effect, it would be legitimate for them to compete for and occupy statuses in the opportunity and power structures of Indian society and to reap the benefits of power, prestige, education, and wealth that flow from them. Moreover, the already acquired statuses of rich man and educated man would not continue to evoke sanctions from other castes. Such statuses were legitimately part of the status-set of one who already occupied the controlling status of Kshatriya. They were, however, traditionally illegitimate for those who occupied the controlling status of Untouchable.

Second, if the Jatavs were accorded the dominant status of Kshatriya in actual interactions, they would not be subject to exclusion and discrimination in informal and private situations. Other castes would not be able to activate against them a salient status of Untouchable. They would be similar to a Negro who could turn white in the United States and who could thus be free to do all things that whites do without fear of retaliation or discrimination.

The sanctions which the Jatavs evoked by their mobility-oriented behavior were not contradictory to the caste system. These sanctions were, in our terms, evidence of conflict, since they were part of the "self-sealing mechanisms"[12] of the caste system. "Sanskritization results only in *positional changes* in the system and does not lead to any *structural change*" (Srivinas 1966: 7). As a means to social mobility, it is based upon an acceptance, not a rejection, of the caste system.[13] The Jatavs were not attempting to destroy the caste system; rather, they were attempting to rise within it in a valid, though not licit, way. They vividly demonstrated this fact when those of them who were most committed to the social map already outlined rejected a claim of certain other Camars, called Guliya, to be Jatavs and therefore to be Kshatriyas. They said that the Guliyas were really Camars and could not be Jatavs, because Jatavs were neither Camars nor Untouchables.

[12] Bailey means by self-sealing mechanisms "institutions which seal off conflict and which tend to keep behavior in conformity with the logical model" (Bailey 1960: 240 and *passim*).

[13] Cohn has noted the same fact for his upward-striving Camars in Senapur village: "It should be realized that, although the Chamars attack the superior social position of the Thakurs and are actively trying to raise their own status, they are not consciously trying to eliminate the caste system. The Chamars are solely interested in raising their status in the system" (Cohn 1954: 262).

Another incident also attests that Jatavs, in attempting to rise in caste rank, accepted the caste system. During the 1930s a low caste spiritual preceptor *(swami)*, one Achhut Anand, led a movement called the Original Hindu Movement (*Adi Hindu Andolan*[14]). This movement had followers throughout Uttar Pradesh, some of them in Agra where the Swami lived and taught for a time. He was driven, some say stoned, out of Agra by the leaders of the Jatav-Kshatriya movement because of his radical doctrines. The Swami taught that the Scheduled Castes were really the original inhabitants of India and that the caste system had been imposed upon them by foreign conquerors. He also taught that the caste system was wrong, and he advocated the abolition of caste differences and caste endogamy. Moreover, he advocated accepting help from Christian missionaries in the form of education, but not in the form of the baptismal waters of conversion. His Christian sympathies made him an anathema to the Arya Samaj-imitating Jatavs. The Swami's advocacy of caste intermarriage, especially with other low castes, went against the grain of Jatav claims to be a separate, endogomous *jati* in the Kshatriya *varna*. Jatav reaction to the Swami is typified by the following passage:

We do not want to be absorbed into others [castes and religions] and thereby lose our identity as some of our Jatav brothers have done by accepting Christianity, after being tortured by false and proud castes. . . .

Would not India brighten her face if the highest qualities of the Jatav race are given the highest place? (Sagar 1924.)

One result of the Jatavs-as-Kshatriyas movement was the rise of new leaders parallel to the old hereditary caste leaders (*chaud-*

[14] A brief history of this movement can be found in Jigyasu (1960). The extent of this movement, as well as other Scheduled Caste movements of this time, has not as yet been explored in North India. Much relevant information has been gathered and summarized in *Rebellious Prophets* by Stephen Fuchs (1965). However, why all these Northern movements failed to gain the strength they did in South India is a question that remains to be answered. The roots of the Northern movements and their possible relations to those of South India are, as yet, unanswered and significant questions for the social historian.

huries). These "big men" were better equipped to handle the problems which Sanskritization had brought to the Jatavs. Many of them were literate, although not necessarily educated, and many, too, were possessed of a worldly confidence developed in business; such confidence the older leaders lacked. Structurally they had added to their status-set the statuses of rich man (*seth, bohare*), literate man (*parhe-likhe admi*), and businessman (*vyapari*). Such statuses were traditionally restricted to the upper castes and were not part of the Camar status-set. It is this inconsistency within status-sets which appears to be one of the determinants of upward mobility within the caste system.[15]

The "big men" used their influence both within the caste and without it, with other castes and the local administration. They functioned as translators, as problem solvers, and as political entrepreneurs.

As translators these leaders tried to make intelligible and desirable the Sanskritic values, rituals, and behavior patterns which they had absorbed from the Arya Samaj.

The Jatav Men's Association put the greatest emphasis on the propagation and extension of Vedic counsels [*sanskar*] and proper traditions. In fact the basis of all social reforms is the following of the counsels. It is only through proper counsels that man can progress. The Association according to its aims propagated among the public the Vedic ways (Yaadvendu 1942: 139).

Since they were literate, and since some were also interested in politics, they communicated to their caste-mates new values and ideas of a Western type. They were, in a sense, pivots between the inner world of caste and the outer world of society, or in Wolf's (1956) terms, they were "cultural brokers."

As problem solvers, the "big men" tried to unravel many of the troubles that Jatavs had with members of other castes and with

---

[15] This change and, as the following pages describe, those that resulted from it seem to confirm Weiner's hypothesis that ". . . the economic improvement of ritually low status groups is likely to stimulate the political organization of such communities" (Weiner 1963: 70).

representatives of the courts and administration. Their advice was
sought also on problems within the caste.

As political entrepreneurs, the function of the "big men" centered on getting recognition for the Jatav claim to Kshatriya status
and Kshatriya perquisites. Before the representatives of the British
Raj, this took the forms of advice, manipulation, and coercion.
Sometimes upper castes were coerced into different treatment of
the Jatavs through the threat of court action; the Jatav Men's
Association on occasion found this method successful. A Jatav historian of the time writes: "In the villages landlords used to treat
the farmers with all possible atrocities, and the situation is the
same today. . . . Legal actions were taken with the oppressors
and they were made to undergo punishment" (Yaadvendu 1942:
140).

One of the first actions undertaken by the Jatav Men's Association was to lobby for the inclusion of one of their members in the
State Legislative Council. Thus, in 1920, a Bohare Khem Chand
was appointed a member of the Legislative Council of Uttar Pradesh, in which he served two terms. After his appointment, Bohare
Khem Chand proposed in the Council that one member of his
community[16] be appointed to each district board in Uttar Pradesh.
The motion was admitted, and Bohare himself was appointed to
the District Board of Agra from 1922 to 1930. He further proposed
that one member of his community be appointed to every municipal board, town area, and notified area committee in the state of
Uttar Pradesh. The motion was admitted and passed. Bohare was
then appointed to the Municipal Board of Agra, where he served
from 1926 to 1928. He also proposed, in 1926, that Jatav (or Untouchable) students should be given scholarships in all schools.
However, he later withdrew this motion on the assurance that the
government would favorably consider the matter. He also became
a member of the Agra Central Jail Committee, the Excise Committee, and the Housing Committee of the District Board. Of even

[16] The word used in the text is *samuday*. The exact reference is unclear, but that
the Jatav community was meant or included is certain. (See Yaadvendu 1942: 217.)

more significance was the testimony he gave before the Simon
Commission Electoral Committee in 1928.[17]

The structural significance of all these added statuses for a man
whose salient status was Camar was that while it did not seem to
give him much direct power, it did give him both a voice to make
his community's demands known and a position of structural observability over the actions of other castes. He could see both what
decisions were made and how they were made in the various assemblies of which he was a member. Bohare then communicated
the results of his observations to the other "big men" and ultimately to the rest of the caste. Thus, the caste itself became more
aware of its position in the power and opportunity structures.
More importantly, it became aware of political instrumentalities
which might be used to influence a change in caste rank. Communications were formally organized through annual meetings of the
Jatav Men's Association. These meetings were held until 1928,
when the organization was dissolved because of internal factionalism. Before its dissolution, however, the movement had spread
from Agra City to all of Agra district and to neighboring districts.

Education among the Jatavs steadily increased. The first B.A. to
a Jatav was awarded in 1926. By 1939 there was an institute called
the Jatav Men's Educational Institute, which established and kept
under its control five schools for boys and one for girls in Agra district. It also had a library at Raja Mandi in Agra City which had
229 books and subscriptions to 16 newspapers and journals in
Hindu, Urdu, and English (Jatav Men's Educational Institute
1939: *passim*). In 1935 another Jatav girls' school, Lady Kailash
Girls' School, was independently established at Panja Madarsa in
Agra. A hostel for boys at Agra College was established by the Educational Institute in about 1939 (Yaadvendu 1942: 229 and 239),
but it has since died out. The fruits of these investments began to
be seen when more men were able to move into statuses previously

---

[17] The Simon Commission was appointed by the British government in 1927 for
the purpose of making recommendations for the revision of the Morley-Minto reforms of 1919 and for extending self-government in India.

closed to them such as government clerks, tax collectors (*tahsil-dars*), and so forth. One man even became a Deputy Superintendent of Police. These educated men formed a new organization in 1930 known as the Jatav Youth League (Jatav Yuvak Mandal), whose purposes were much the same as the older organizations. These men also urged members of their caste to adopt Vedic (Arya Samajist) ways and ritual. From time to time, newspapers were published to spread this message as well as more general information useful to the caste. The growing political consciousness of the caste was evident in the annual resolutions of the Jatav Youth League. By 1941, it had branches in other states such as Rajasthan, the Punjab, and the old state of Madhya Bharat. The following resolutions were passed in 1937 and 1938, respectively:

The conference takes a decision that the aim of the League in the future will be to attain political freedom by suitable constitutional and peaceful means.

This conference would impress upon the Jatav members of the Provincial Legislative Assemblies that regardless of the political party to which they personally belong, they should vote unanimously on any issue concerning the interests of their community (Yaadvendu 1942: 163).

The anomalous fact of Jatavs occupying statuses such as tax collector and Deputy Superintendent of Police increased the structural pressure upon the local caste hierarchy to accord to the Jatavs a controlling status of Kshatriya which would be concordant with their new statuses. It also put faint pressure upon other castes to accord Jatavs a dominant caste status of Kshatriya in actual interactions. To reiterate a point, Sanskritization was the traditional "self-sealing mechanism" of Indian society for dealing with such situations of status inconsistency and conflict. It should also be noted once again, however, that the Jatavs-as-Kshatriyas movement was not against the caste system as such. The goal of the Jatavs was merely to raise themselves in the caste hierarchy by changing their *varna* position. This is clear from the Jatav Youth League's resolution of 1939 against the Original Hindu Movement, which had as its expressed aim the end of the caste system and caste endogamy.

> This conference declares that Jatavs have no connection with the "Original Hindu Movement" and considers the Original Hindu Movement to be harmful for the Jatav race. Therefore, the workers of the League, the members of the League, and the Jatav people should not take part in the Original Hindu Movement otherwise the League will take disciplinary action against them (Yaadvendu 1942: 167).

During the 1930s, when the Jatav Youth League was organized, the Round Table Conferences[18] were taking place in England. As far as the Scheduled Castes were concerned they had two protagonists, both of whom claimed to be "the" leader of the Untouchables: Mahatma Gandhi and Dr. B. R. Ambedkar. Of the two, only Ambedkar was an Untouchable. He was a man of exceptional qualifications and accomplishments. The struggle between Ambedkar and Gandhi over the issue of separate elections for Untouchables was known to the Agra Jatav "big men."[19] They, along with other Untouchable groups, wired London to insist that Ambedkar, not Gandhi, was their leader. The political importance of this telegram was tremendous. It signaled that Jatavs had become aware of the national implications of their problems and activities through a process of analogical identification. That is, their problems, insofar as they were matters peculiar to Agra, were local and unique; but insofar as they were a matter of Untouchability, the problems were national and general. This important process continues to operate even today. Through it, the narrow boundaries and primordial loyalties of caste are transcended, and a unity of disparate castes in pursuit of common goals, even at the national level, is achieved. This is especially so in the realm of politics.

Even as early as the 1920s, the caste gave a public demonstration

[18] The Round Table Conferences were held in England during 1929–32. Their purpose was to sound out Indian opinion on the nature of reforms for the proposed new constitution of India, subsequently known as the India Act of 1935.

[19] This event is discussed at greater length in Chapter V.

of its offensive stand. At that time a certain Jatav named Karan Singh sold snacks in Belanganj, Agra. A high caste person bought and ate, unknowingly, fried food from Karan Singh's hand. In India food cooked by a lower caste person is polluting to an upper caste person; especially is it so when the cook is an Untouchable. When the fact of the seller's caste became known there was a quarrel, and in the melee Karan Singh was beaten and died from the blows. The Jatavs organized a procession and angrily paraded through Belanganj. In an act of defiance they took out a bamboo pipe from the water stand (*piao*) there. The bamboo pipe was for the use of the Scheduled Castes only; they were too polluted to receive a drink of water directly from the spigoted brass pot (*sakha*) meant for the touchable castes. After that, a law suit against the alleged killers was instituted and financed from donations by Jatavs. Karan Singh is now proudly remembered as the first Jatav martyr (*shahid*).

Further political action by Jatav "big men" during the 1930s involved lobbying for the acceptance of the name, Jatav, by the provincial and central governments. The first attempt to have the caste listed separately in the Census was made in 1931 in the form of a letter to the Viceroy and the Census Commissioner. Later, when the Jatavs were not listed separately in the 1936 list of Scheduled Castes, the movement went into full operation with three demands: (1) that in all government papers the caste be listed as Jatav and not as part of the Camar group of castes; (2) that the caste be listed separately as Jatav in the Census; and (3) that the Uttar Pradesh government recommend to the British government the acceptance of the Jatav community as a separate caste in the list of Scheduled Castes. Once again the Jatavs took the position that they were not Camars and that they should not in any way be included among them. In a letter addressed to the Secretary of State for India in London, dated December 1938, they took the position that Jatavs were in a depressed condition only because of the accidents of history and, therefore, they were not really like other Scheduled Castes or Untouchables. The letter says:

It is needless to mention that the Jatavs, as constituted at present, form the most prominent class of the community styled as Scheduled Castes. . . .

We desire to make it plain that we, as Jatavs, claim to be recognized as a separate caste amongst the Scheduled Castes without being amalgamated with other castes under the list of Scheduled Castes with which we have no endogamous connection (All-India Jatav Youth League 1938).

Their objective was to have the government recognize the old claim that the Jatavs were really Yadavs and therefore Kshatriyas. By this time they were even able to quote British authorities to support their claim; they brought in Nesfield's speculation that Jatavs were possibly an offshoot of the Yadu tribe from which Lord Krishna came.[20] If this goal were achieved, then their claim to Kshatriya status would at least be recognized by the government.

The political maneuvers undertaken in the push for recognition of their status were not unsophisticated for a caste which previously had been totally depressed and powerless. A deputation of Jatav "big men" put their claims before Govind Vallabhai Pant, then premier of the Uttar Pradesh government. One of the two Jatav members of the Uttar Pradesh Legislative Assembly (elected in 1937) questioned in the Assembly the Uttar Pradesh Minister of Education on the same problem. In addition, a memorial letter, said to bear six thousand Jatav signatures, was sent to the Secretary of State for India in London. As a result all the demands were accepted by the government. Unfortunately, World War II and the independence movement intervened, so that these decisions were never followed through.

Nevertheless, the conclusion that the Jatavs had a minimally effective organization for interest articulation and political action

---

[20] "Jatav is an occupational branch of the Yaduvansh in which Shri Krishna was born" (quoted in Yaadvendu 1942: 96; see also Briggs 1920: 23). The textual evidence in my possession makes it difficult to decide whether this idea was first expounded by Nesfield or by Swami Atma Ram in The Lomash Ramayana. In any case, there seems to be some mutual influence.

is inescapable. The structural fact on which these actions were based is the expansion of the Jatav status-sets, which by 1942 included educated men with college degrees, Members of the Legislative Assembly, tax collectors, inspectors of police, and government clerks. The men who occupied these previously unattainable statuses did as much as they could to strengthen the Jatav foothold in them. The fact remained, however, that the Jatavs themselves were unwilling to grant these statuses to other Untouchables in view of their own claim to Kshatriya caste status.

In this fluid situation, three significant anomalies remained. The first was that the occupation of shoe maker, upon which the Jatavs based much of their economic independence, was still part of their status-set. As such, it had high visibility and thus easily identified them as Camar and, therefore, as Untouchable. Were they able to leave this polluting occupation, their claim to a new dominant status might more easily have been achieved and legitimized. The Jatavs were not trying to add new statuses to the mutually exclusive status-set of Camar; rather, they wanted to supplant the Camar status-set with the mutually exclusive status-set of Kshatriya. They could not easily do this as long as they retained the highly visible and still stigmatizing occupation of shoe maker or Camar.

The second anomaly was that the Jatavs freely chose, in fact fought, to be included on the list of Scheduled Castes. This option brought them concrete benefits from the government, but is also continued to identify them as an Untouchable caste, a position quite inconsistent with their claim to Kshatriya status.

The final anomaly was the importance to the Jatavs of their negative reference group, the Brahmans. While on the one hand they resented, even hated, them, on the other hand Jatavs needed them and curried their favor. The British might recognize and accept the Jatavs as Kshatriyas, but the Brahmans alone could lend ritual legitimation to the Jatav claim by recognizing it and agreeing to officiate at Jatav ceremonies. Even the Arya Samaj seemed to have failed the Jatavs here. One informant told me, from the Jatav point of view, a story of how the Arya Samajis were invited to a special feast given by the Jatavs. When the appointed hour came, few Brahmans showed up and even fewer ate with gusto. Arya Samaj acts seem to have belied their words. Accordingly Jatavs were left with few illusions about the Samaj.[21] Disillusioned or not, however, the necessity for continued Sanskritization on the part of the Jatavs remained, because without it no hope of Brahman legitimation was possible.[22] Indeed, it was probably a vain hope anyway, since it is not known that Sanskritization has ever worked for an Untouchable caste.[23]

One further point must be noted. None of these Jatav actions can be considered contradiction as I have defined it; on the contrary, it was conflict within the caste system. Calling upon the British to validate their asserted identity is in accord with traditional "self-sealing mechanisms" for dealing with conflict within the caste system. Such validation was formerly a prerogative of the king, whom the British had supplanted. The Jatavs, unlike the Original Hindus, were attempting to rise within the caste system and thereby accepted it. The intended consequence of their social action, at that time, was not to destroy the caste system; it was only to rise within it.

[21] This is not to say that there were not many sincere members of the Arya Samaj. I myself have observed one such venerable old man at a Jatav wedding. Jatav food seemed no problem for him.

[22] "Ambitious castes were aware of the legitimizing role of the Brahmin . . . even the powerful rulers of the Vijayanagar Kingdom (1336–1565) in South India had to acknowledge and pay a price for the legitimizing role of the Brahmin" (Srinivas 1966: 28).

[23] This is Srinivas's (1962: 58–59) view, and I tend to agree with him. The problem of the tribals who are sometimes assigned Untouchable status by caste Hindus is a complex one and brings up the problem of the many meanings of Sanskritization. Some tribals who enter the caste system enter at a high rank, for example, the Coorgs who entered the system as Kshatriyas (Srinivas 1952) through Sanskritization. However, entering the system and moving up in it are two very different things. If a tribal has entered the system *unambiguously* as Untouchable and is no longer thought of as a tribal in any way, then I doubt that Sanskritization would work as a means of rising in the system.

Transitional Period

During the 1930s and early 1940s, the Jatav elite turned its attention to new problems. These were the independence movement and the position of the Jatavs in an independent India, although the latter issue was overshadowed to some extent by World War II. In this shift of attention three things happened. First, the more liberal notions of the independence leaders, such as democracy, individual freedom, and individual achievement, echoed mightily in Jatav ears (the last idea was already familiar to them through the Arya Samaj). Second, the notion that India was to be free and independent was interpreted: "If India is to be free, then the Jatavs, too, must be free within an independent India." And third, the undercurrent of opinion, which favored the views of Dr. Ambedkar and which was foreshadowed in the telegram sent to London in his support, burst forth and became a major point of view among Jatavs.

In 1944–45 these political undercurrents crystallized into the formation of the Scheduled Castes Federation of Agra which was linked to Dr. Ambedkar's All India Scheduled Castes Federation. It was a major turning point in the Jatavs' definition of their social situation. Consequently, their reference group of identification now became the Scheduled Castes with whom they identified as the oppressed, unenlightened, and deprived section of the population. These characteristics of the Scheduled Castes became the basis of a new Jatav self image; it was a striking about face from their march to Kshatriyahood. The Jatavs now saw *themselves* as one among the castes considered Untouchable by the *others,* the touchable castes.

This about-face was influenced by a concurrent change in the Jatav reference group of imitation. This group included the liberal leaders of the independence movement in India and, for some of the Jatav elite, leaders of other independence movements among oppressed peoples such as the French and American Revolutions. They formed the activist group to be imitated in the struggle for

the liberation of the oppressed and the poor. They were the ideal-
ists who pointed the way to the utopia that independence was to
usher into India.

The reference group of negation remained the Brahmans, who
continued to be seen as cruel demagogues. In Jatav eyes, it was the
Brahmans who were torturing the Scheduled Castes on the rack
of Untouchability and the caste system. These three reference
groups—the Scheduled Castes, the independence leaders, and the
oppressing Brahmans—redefined the Jatav position in Indian society at that time.

The Jatavs thereafter became definitely antagonistic to casteism
and the caste system. Much of this was due to the influence of
Ambedkar, who was both the formulator and the leader of the
new movement (see Chapter VI). The Jatav elite was cognizant
of his ideas and his objectives, which he summed up in the historic
phrase, "Liberty, Equality, and Fraternity." Ambedkar's experience as a student in America, where he took a Ph.D. in economics,
and in England, where he became a barrister, no doubt colored
the interpretations he gave to this phrase in India.

The years of "political socialization" that had preceded this redefinition thereafter became years of "political recruitment," to
use Almond's phrases (1960: 17ff). A united Jatav front was
formed. Leadership moved steadily from indirect political influence to direct political participation. This shift was solidified by
three acts. The first was the act of civil disobedience (*satyagraha*)
at Lucknow by the Scheduled Caste Federation, in which the Agra
Jatavs claim to have taken an important part in terms of both
numbers and leadership. According to the Agra organizer of the
act of civil disobedience,

The reason for the movement was that we felt the British Raj was
coming to an end and that if Congress came into control we would not
have any rights because of the Congress being the party of the caste
Hindus. . . . It was this movement that was the real awakening of the
Scheduled Castes. From this time on, they were really conscious of the
battle that had to be fought.

The leaders of the movement presented eleven demands to the provincial government in Lucknow, including one for the reservation of government jobs for the Scheduled Castes.[24] Participation in the act of civil disobedience is still a mark of honor for those Jatavs who can claim to have marched or to have been imprisoned because of it. Clearly, Gandhi's technique for winning independence was imitated and turned into a technique for getting the "rights" of the Scheduled Castes. Indeed, the use of this powerful political tactic is not forgotten to this day.

The second aggressively defiant act was a parade through the streets of Agra. In it the main attraction was a wolf symbolizing the Poona Pact.[25] The Poona Pact was like a Braham wolf that had run wild and was devouring the Scheduled Castes. Ambedkar interpreted the Pact as a betrayal of the Scheduled Castes by Gandhi, an interpretation which was and still is accepted by the Jatavs. The wolf was publicly burned, and the heat of Jatav anger was not lost to the eyes of the Agra upper castes who saw themselves symbolically consumed in the flames of long smoldering resentment.

The third event of this period was the election of 1946 to fill the reserved seats in the Legislative Assembly. It was this election

---

[24] A leading English language newspaper of Lucknow, the *National Herald,* gave little publicity to this act of civil disobedience except to note in inner pages that groups of 60–70 and so forth, were put in prison for defying Section 144 Cr. P.C. Most of these notices were of about four or five lines and none mentions the aims of the movement. This news blackout seems significant. It seems especially so considering that a fairly reliable recorder of the movement told me that a total of 3,023 persons were, at one time or another, in prison; that more prominent notice was given to the activities of Mr. Jagjivan Ram, the Congress Scheduled Caste leader; and, that allegations of Muslim League support were made against the act of civil disobedience.

[25] This event is treated at length in Chapter V. Briefly, the Poona Pact was an agreement between Dr. Ambedkar and Gandhi over how the Scheduled Castes were to vote in elections. Ambedkar wanted separate electorates for the Untouchables. At the knife point of Gandhi's fast unto death over this issue, Ambedkar capitulated and gave up separate electorates for joint electorates in return for an increased number of seats in the Assemblies. These seats are reserved for members of the Scheduled Castes alone, although they are elected by both Scheduled and non-Scheduled Castes. All other seats are known as general seats, and both Scheduled Castes and non-Scheduled Castes can contest elections to them.

which split the united Jatav front in two. The split ran between the conservative followers of the Congress Party and the revolutionary followers of the Scheduled Castes Federation. Originally, the conservatives ran things, but they have steadily decreased in popularity to the point that today the more liberal Jatavs are in command. In the 1946 election the Jatav Scheduled Castes Federation candidate, who was a former Congressman and M.L.C. (Member of the Legislative Assembly), was defeated by a Jatav Congress candidate. The Congressman won, it was said, because of the joint electoral system in which upper castes also voted for him as the Congress candidate from the Scheduled Castes. No doubt, too, the prestige of the Congress Party was still high at the time and contributed greatly to the success of its candidate. Conservatives were successful again in 1952 and 1957. The defeat of the Scheduled Castes Federation candidate was undoubtedly due in part to its own internal factions, lack of organization, and lack of money.

Post-independence Period

For the Jatavs, the transitional period quickly came to a close with the achievement of independence in India. This historic event brought with it parliamentary democracy and a constitution whose basic principles included democracy, freedom of the individual, and, more importantly, the establishment of the universal franchise in India. This new dispensation created a basic change in the status-set of the Jatavs. They henceforth were both citizens, equal to all other individuals, and voters equipped with a powerful tool, the vote. In the words of one Jatav politician:

The Jatav may not understand the intricacies of politics but he does understand the vote. The vote means he can get [*sic*], sometimes by selling his vote, or by electing the right man. He understands the power of the vote and wants to use it.

And another: "The ordinary man does not understand the intricacies of politics; neither do I. But he does understand the simple things about politics and that it is a way of getting his rights."

By exercising the rights given them in these new statuses, the Jatavs are now trying to gain further entry into the opportunity and power structures of India. Through them, also, they are trying to make the politically ascribed status of citizen, and not the religiously ascribed status of caste, the dominant status in cases of conflict or contradiction between interacting Indians. The social situation is one in which both caste and class norms now exist.[26]

Article 17 of the Constitution added to these changes the abolition of Untouchability and forms of public discrimination. Two further pieces of legislation, the Acts of 1955–56 unifying Hindu laws on marriage, inheritance, and so forth, and the Untouchability Offenses Act of 1955, further strengthened the position and safeguarded the rights of the Scheduled Castes.[27] In effect, these pieces of legislation made caste illegal as a dominant or controlling status and replaced it with the status of citizen. Jatavs would often point out to me that all men are Camars (leather workers), because all are made of skin, and therefore all are Camars by nature. Others told me they no longer considered themselves Scheduled Castes, rather they were human beings or members of mankind (*insan*).

The appointment of a Commissioner for Scheduled Castes and Tribes brought in the state as a third party to safeguard the rights of the Scheduled Castes. This third party at the local level is the Harijan Welfare Officer. By definition he is not neutral between the statuses of caste and citizen and the systems in which they operate. Contradiction is therefore now possible since he, as the representative of the state, is legally bound to work in favor of members of the Scheduled Castes insofar as they are equal citizens, though informally caste inequalities may continue to oper-

[26] This conflict of caste and class norms also exists in villages of India. The whole subject is incisively discussed in Beteille's (1965a) book, *Class, Caste, and Power*.
[27] This legislation and the problems involved in it are admirably summed up in Galanter (1963).

ate.[28] That they do continue to operate was poignantly demonstrated to me by the Harijan Welfare Officer in an incident I
should like to parenthetically relate here. It seems that the government had requested the Harijan Welfare Officer to get together
a program advocating the elimination of Untouchability. This
was to be given in remembrance of Gandhi on his birthday. The
program of speeches with tea and biscuits afterwards was given
in Bhim Nagar, a Camar neighborhood of Agra. After the program the Welfare Officer admitted that Camars had little need to
be exhorted to give up the practice of Untouchability. Almost
apologetically he said that the program should have been given in
a high caste neighborhood, but such a neighborhood would have
none of it. One Camar said that the Harijan Welfare Officer had
to fill out his forms noting that the program was appropriately
conducted, so the people of Bhim Nagar obliged him by allowing
it to be celebrated there.

Along with these changes has come the possibility of "bridge
actions" [29] within the Indian social system. Thus, an individual
may activate in particular situations either his caste or his citizenship status, whichever is convenient for obtaining his goals. What
has happened, then, is that caste status has to some extent become
"politicized," in the sense I have defined the term. Contrary to
popular opinion, including popular Indian opinion, castes have
not been legally abolished in India; they still remain as legal
entities.[30] Indian law has only, and realistically, tried to abolish
caste inequalities and discrimination insofar as it is a matter of
public concern. Thus, the government by taking an interest in

[28] It is a constant complaint of the Jatavs, especially of those who are members
of the Republican Party, that the Harijan Welfare Officer really does not want to
assist them, since he is an upper caste man. They want a Scheduled Caste person
to hold this office. In other words, they feel that while publicly he accords them
the dominant status of citizen, nevertheless privately he activates their salient
status of Untouchable. I was not able to find convincing proof of this Jatav complaint.

[29] ". . . the actor may play upon the roles which he has in different systems of
social relationships, so as to win for himself the support of more effective allies"
(Bailey 1960: 251).

[30] For a lawyer's explanation of this, see Galanter (1961 and 1963).

the Untouchables has redefined by dichotomization their caste status. For specific public purposes they are Scheduled Castes, while for other purposes they remain Camars, sweepers (Bhangis), or whatever. The government's special role exists in its welfare programs and its programs to safeguard the rights of the Scheduled Castes. Thus, the Scheduled Castes have to some extent been integrated into the nation and state at levels beginning from the Office of the Commissioner for the Scheduled Castes and Scheduled Tribes in New Delhi down to the Harijan Welfare Office in Agra. Moreover, other castes, in their roles vis-à-vis the Scheduled Castes, are also subject to the scrutiny of the government, to the extent that their caste status has also been "politicized" and made subject to standards of behavior other than those based on caste.

These changes in the structure of Indian society were basic to the redefinition of the social situation made by the Agra Jatavs in independent India. First of all, in 1956, most of the Agra Jatavs followed Dr. Ambedkar into Buddhism. As Buddhists, they claim to be the original residents of India, who were forced into servitude by the Brahmans and the instrument of their oppression, the caste system. This again is a variation on the "original Indians" theme which had already appeared in the Original Hindu and Jatavs-as-Kshatriyas movements. Identification with the ancient Buddhists was more than a religious move. I shall discuss its many ramifications in the next chapter; but for the moment I briefly mention four important elements involved in the conversion to Buddhism.

First, Buddhism was an indigenous Indian religion and therefore could not be treated as foreign and suspect as could Christianity and Islam. Second, it was strongly anti-caste, at least in Ambedkar's version of it.[31] Thus, it appealed to an indigenous but non-Hindu tradition. Buddhism presented an alternative to the caste system in which Jatav attempts at social mobility had

---

[31] Ancient Buddhism's exact opinion on the caste system is ambiguous and a subject of scholarly controversy. However, for an impassioned polemic against caste, especially the Brahmans, the *Vajrasuci* of Asvaghosa (1950) has few equals.

been unsuccessful. It can be seen, then, as an adaptive response on the part of many Jatavs to the emerging social structure of post-independence India. It was also an attempt to fill the vacuum created by the loss of the unifying ideology of the independence movement. Mahar has very perceptively pointed out this vacuum in her study of Khalapur village:

Abolition of untouchability, and social and economic advancement for untouchables, are justified by the Merchant, the Sweeper, and the Chamar in terms of the goal of independent rule for India. This goal was achieved more than a decade ago and as the Chamar points out, some of the immediate enthusiasm for equality has been dissipated. These men do not justify equality in terms of the need to maximize talents and ability for India's economic development, a rationale offered in India's Five-Year Plans. Nor do they discuss the dignity of the individual as does the Preamble of the Indian Constitution. It appears that if Khalapur is to accept the value of personal equality, some new set of justifications for caste equality must be introduced (Mahar 1958: 63).

Third, because Buddhism exists in countries outside India, the expectation that non-Indian Buddhists would take the cause of the depressed and "persecuted" Buddhists in India to an international forum was voiced by a number of my informants.

And finally, the "We are the original Indians" theme gave an ideological and a moral justification to Jatav political demands for "giving the land back to the tillers and the government back to the people."

A second element necessary for the Jatav redefinition of the social situation was to find a negative reference group; this was and still is the Brahmans. But today it is the Congress and the Jan Sangh Parties which have come to represent the Brahmans. Moreover, in Jatav eyes, these have come to include the capitalists (*punjipatis*) such as Birla and Tata and the rich men (*bare petvalas*), a caste-anomalous class of the bourgeoisie. In Jatav opinion,[32]

---

[32] Cohn (1954: 171–72) notes that the Camars of Senapur village have also become disillusioned with the Congress and call it a bad party. Yet no other party has come to take its place.

The political parties other than the Congress which have been established in India are organized and conducted either by the big men or
by the "Brahman *Gurus*" [teachers]. The Congress, which is in power
in the country, is an institution conducted, directed, and led by the big
men. . . . Furthermore, the Hindu Mahasabha, the Jan Sangh, the
Ram Rajya Parishad, the Varanashram Swarajya Sangh, etc., have been
organized by the "Brahman *Gurus*." Their aims are to establish Hindu
rule and Brahman command; therefore they are considered communalistic organizations which are far from democratic (Jigyaasu n.d.: 4).

The lack of differentiation between religious caste status and
secular class status evident in this passage indicates that both caste
and class attributes operate as the bases of power in changing
Indian society.

In Agra, the Punjabi factors also form part of the negative
reference group against whom the Jatavs are pitted. These factors
are resented, because they take the major share of the profits in
the shoe trade and because they are now trying to set up or buy
shoe factories of their own. This is seen as an invasion of the exclusive ownership of the means of production which Jatav's had
for some time. The situation is aggravated, because the Punjabis
have capital while the Jatavs have little or none. A Jatav writing
in a local newspaper describes, with pinpoint accuracy and no
little sarcasm, how Jatavs view the situation. He writes:

> Formerly we were slaves of the Mohammedans. We enjoyed great
> profit by wearing felt caps for some time. Our brothers who have come
> from the Punjab have obligated us to them in one way; they have
> freed us from the slavery of the Muslims and put us in their own
> slavery. They have provided for many of our brothers by opening shoe
> factories in Agra and Delhi.
>   The day is not far away when not a single Jatav factory owner will
> be visible, and all the factory owners will have turned into shoe makers.
> . . . Have the Jatav factory owners seriously considered the conse-
> quences of the policy they are following? (*Nau Jaagriti*, October 22,
> 1956.)

In this passage the acid of sarcasm burns deepest in the word
"brothers."

In short, the relationship of the Jatavs to their negative reference group is one of relative deprivation. They feel that the Punjabis, the Brahmans, and the capitalists hold an unmerited share of
Indian gold and Indian soil, which they have obtained through
exploitation of innocent Untouchables. They feel too that the
Brahmans and their cohorts have made Indian politics and administration into an oligarchy through the ritualistic chicanery of
the pollution concept. The independence movement, schools, motion pictures, political rhetoric, Bhavean reformers, government
leaflets, and cosmopolitan urban life have acutely sensitized and
increased Jatav awareness of deprivation and new norms of social
justice. (See Tangri 1962: 208.)

The third element in the redefinition of the social situation was
the formation of the Republican Party of Agra in 1958. It was
formed as a branch of the All-India Republican Party and as the
successor of the Scheduled Castes Federation. Other parties of
India, especially those in opposition to the Congress Party, were
the reference group of imitation upon which this organization was
anchored. For the Jatavs of Uttar Pradesh and the Mahars of
Maharashtra, both of whom were followers of Ambedkar, the
formation of the Republican Party was an adaptive response to a
political system in which their interests were not being realized
nor articulated to the extent that they desired. Structurally, it
was and is an organization based on the statuses of citizen and
voter in a parliamentary democracy.

The Republican Party and the Buddhist movement are in a
complementary relationship. The Party emphasizes the "exoteric" [33] issues, the harsh here and now, the economic, political,
and social plight of the Jatavs as they live with the other castes;
while the Buddhist movement emphasizes the "esoteric" issues, the
"sweet bye and bye," the millennial future when the Buddhists
shall inherit the land and caste will no longer exist. Buddhism

[33] The distinction between esoteric and exoteric teachings was taken from
Essien-Udom (1964: 22).

provides a psychological justification and an ideological rationalization for political militancy.[34]

The relationship of the Party to the Buddhist movement is close, because Dr. Ambedkar, in effect, founded both. Much of the Republican Party literature is concerned with the fact that Dr. Ambedkar, the apotheosized (as a Bodhisatva) culture hero of the Jatavs, was the major architect of the Constitution of India; and not without accident are many of the ideals of the Buddhists also found in the Constitution of India. Moreover, there is much truth in an informant's statement that "Whoever fights an election in the name of Ambedkar wins." For these reasons, there is a strong attachment to the Constitution and the parliamentary system as envisioned by Ambedkar, who is affectionately called Baba Sahab or Respected Sire.

Baba Sahab framed the Constitution. . . . Baba Sahab placed in it the highest principles of justice, equality, freedom and fraternity. If the 90 per cent of the people in India who are oppressed, backward, exploited, and deprived want to take these rights by constitutional means, they can do so. And if they want, they can also acquire control of the whole administrative machinery. The Republican Party of India is the most valuable and proper means for achieving the objectives of the Constitution (Jigyaasu n.d.: 7).

In such a context, it was a stroke of genius, whether conscious or not, for Ambedkar to define the roles of Buddhist and citizen in

[34] Commenting on Gough's (1955) study of a Tanjore village where the Untouchables are losing interest in the large temple festivals which were opened to them in 1957, Dumont and Pocock (1957: 38) note: "This section is in fact dual, it is against the dominant caste as a political power and against Brahmanism because of its association with dominance. We are told that the Brahmans are for the most part voters for the Congress Party while the Adi Dravidas are increasingly influenced by Communism. Imitation of any superior custom—religious or secular— in these circumstances is an admission of inferiority incompatible with the political aspirations of the lower castes."

If the Brahmanism and the superior customs in the above quotation are read as the negative reference group, and Communism is read as the Republican Party, or reference group of imitation, the situation in the South can be interpreted to be the same as it is in Agra. The Jatavs, too, reject superior customs and inferior status and identify themselves as Buddhists, who in secular terms identify themselves as equal human beings or citizens.

virtually the same terms. The interchangeability of these two statuses allows for an appeal to the religious and traditional-minded as well as to the revolutionary and secular-minded; yet, both can work for the same revolutionary goals.

Under new conditions of socio-political structure in which the statuses of citizen and voter cross-cut both caste and class, it is easy to see why the Jatavs have given up claims to a dominant status of Kshatriya, and efforts to imitate Sanskritic cultural behavior. It is also easy to see why they have become antagonistic to casteism and the caste system. In effect, they have now reversed their old stand against the Original Hindu Movement. The change is due to the fact that Sanskritization is no longer as effective a means as is political participation for achieving a change in style of life and a rise in the Indian social system, now composed of both caste and class elements. The ultimate object of Sanskritization was to open and legitimize a place in the opportunity and power structures of the caste society. It is hoped that the same objectives can now be achieved by active, but separatist, political participation. No longer is entrance into the power, opportunity, and wealth structures of Indian society based solely on ascribed caste status; in the public sector, at least, achievement based on citizenship status is the principle of recruitment to these structures.[35]

The Republican Party is the organized response of the majority of Jatavs to gain these ends. Caste-based inequality and discrimination are now illegitimate, though they do continue to function latently and, no doubt, efficiently (Gould 1963). The upper castes now resent and label the Jatavs as troublemakers and non-conformists. Yet in this case it is the rebellious minority that has grasped the thread of the ultimate values of the Indian republic far more readily than the conservative majority.

[35] Long ago Bailey (1957: 227) noted that the Board outcastes of Bisipara village in Orissa were tending to separate themselves from the rest of the village and to order their relationships with other castes in the village not through the village structure but through the administration. The reason is similar if not identical with that given above. See also Rowe (1960: 58) for a similar movement of the Noniya caste in Uttar Pradesh, which abandoned the Arya Samaj for direct political action.

It is easy to understand that the switch from Sanskritization to
direct and overt political participation occurred, because in independent India the latter was more useful for achieving access to
the strategic resources of wealth, power, occupation, and education. However, this does not explain the identity transformation
into Buddhists. The answer to this problem, I believe, is that Buddhism offered a resolution to the state of "cognitive dissonance"
(Festinger 1957), in which the Jatavs found themselves. The Jatavs
sought to be mobile and to improve their condition. But to Sanskritize would mean that they accepted the caste system and caste
inequalities. This was logically contradictory to their acceptance
of democratic equality and citizenship for all, and their goal of
eliminating the caste system. Buddhism was truly Indian, yet it
was also ideologically consistent with their goal of mobility and
the new ideas they had come to accept. Buddhism, then, redefined
the Jatavs' map of social space and their aggressive mobility-
oriented strategy in it. It did so, however, in a culturally relevant
idiom and ideology that was consistent with the principles of the
new constitution which the Jatavs espoused and sought to use to
their advantage.[36] This was all that the Sanskritic ideology was not.

Party Politics

The Republican Party is for all practical purposes an adaptation
of caste organization to the changing social and political structure
of Indian society. The Party performs four functions for its members: these pertain to organization and development of power,
policy, leadership, and integration.

Through the Republican Party the Jatavs attempt to place their
own men in status positions that will help them to hold power
and bolster their position in the economic and opportunity structures. In democratic India, they now have the opportunity to do
this through the statuses of citizen and voter. Thus, elections to
legislative posts have assumed great significance; they are the

---

[36] For a further elaboration and development of this point, see Lynch in *Untouchables in Contemporary India,* edited by J. Michael Mahar (in press).

At left, a Republican Party parade showing Ambedkar's picture and
the party flag. Right, the newly elected (October 1963) president
of the Agra Republican Party.

structural and legitimate means with which to capture power, at least the power of authority,[37] in Indian society.

In 1959 Agra City became a municipal corporation under the new Municipal Corporation Act of the state of Uttar Pradesh. This same year, elections were held for fifty-four members of the Municipal Corporation. The Republican Party won seventeen seats, six of which were reserved for the Scheduled Castes. Of these seventeen, eleven were Jatavs. Four of the seventeen, all non-Jatavs, were supported by the Party, but soon after the election, they severed any connection with the Party, thus bringing the Republican total down to thirteen members. There were also three Independent Jatav candidates who periodically associated themselves with the Party and derived support from it. Thus, there were sixteen effective members of a Republican-Jatav coalition in the Corporation. To this number one other Jatav was co-opted, bringing membership of the coalition to seventeen. One member of this total number is a Mohammedan, now only peripherally associated with the group, and one other is a member of a merchant caste or Vaisya, leaving fifteen Jatavs in the coalition.

Except for the Vaisya, who is a medical doctor, the only highly educated member of the Republican Party corporators is the man who was co-opted. After his co-option into the Corporation, he was also elected to the office of Member of the Legislative Assembly of Uttar Pradesh from a general, not a reserved, seat.[38] Because he is an avowed Buddhist, he was ineligible to stand for a reserved seat. The success of his election from a general seat is considered a great victory for the Party and the caste. This man is now studying law because he feels that with a law degree he can more effectively serve his community in the Legislature.[39]

There are also two other Jatav members of the Legislative

[37] I follow here Weber's distinction between power and authority. Authority is legitimated power vested in a particular status or society. Power is the ability to force or compel others to do one's will; it is not necessarily legitimated by a particular society (see Weber 1958: 100 and 295).

[38] For the distinction between general and reserved seat, see Chapter II.

[39] There is only one Jatav lawyer in Agra today, although there are a number of law students. He, alas, is not a very good lawyer, as the Jatavs themselves admit.

Assembly resident in Agra, bringing the total to three. One, who is also a municipal corporator of Agra City and intermittently a Republican, was elected an M.L.A. from Fatehabad district from a reserved seat. The second was elected from a reserved seat in Agra on the Congress ticket. He received much of his support from the Jatav Congress "big men." He won the election by 1,047 votes over the Jatav Republican candidate's 20,865. In this defeat of the Republican candidate, three factors played a role. First, a third Jatav entered the contest, because, he says, he had a grudge against the Republican candidate and wanted to defeat him. This he did by capturing 2,852 votes. Second, the constituency is a rural one, and the Republicans are not as effectively organized in the rural areas as they are in the city. And third, the Congress attempted to activate as salient the Untouchable caste status of the Republican Jatavs in the election.

The third seat was a reserved seat and from it a Congress candidate from the Scheduled Castes won because the caste [*svaran*] Hindus felt that they could defeat the Republican Party's Jatav candidate by making the Congress Jatav candidate win. Like all other parties, the Congress candidates did not get even one per cent of the votes in the name of socialism or informed opinion. Of the votes which were won, 70 per cent were gotten in the form of influence [*satta*], money and communalism or anti-communalism. The primary and most effective reason for the success of the two Congress candidates for the Legislative Assembly was that nowhere was there more evident the feeling of the communalism of the caste Hindus than against the Republican Party candidates (*Sainik* 14 May, 1962).

Significantly, this passage was written by a well-informed and experienced Brahman politician who has recently joined the Congress. He can hardly be accused of bias against the upper castes.

Some of the success the Jatavs had in these two elections may be due to a number of factors peculiar to Agra. The Jatavs form about one-sixth of the city's population. Moreover, they are concentrated into segregated caste wards of the city. These two factors, numbers and segregation, are sociological facts which, given an enlightened electorate, are powerfully adaptive in a political system with electoral wards and the numerical vote. Added to this was the experience gained during the long period of political socialization and enlightenment (*jagriti*) during pre-independence years. The Jatavs had then and have now a literate elite, and to some extent a literate public, which continues to make effective use of the communications media, such as newspapers and leaflets, available in the city.

In the 1962 election, there was a union between the Muslims and the Republican Party. A Muslim was nominated for the Parliament seat and three Republicans, including the Vaisya doctor, ran for the State Legislature. The Jatavs as well as the other castes can activate caste as salient in any situation in which it is useful for them, although it is not necessarily the status which is formally appropriate to the situation. I wish to emphasize the point that while caste as a salient status is no longer legitimate and manifest, it does continue to function latently. Thus, after this election two leaders of the Republican Party published a public denunciation of a slogan which had been raised during the campaigns. The slogan implicitly accused the Jatavs of an appeal to communalism. The Republican leaders wrote that it was not true that they had used (although they probably made no effort to suppress) a slogan appealing to communal loyalties. They denied that the Republican Party in the last election used such a slogan as,

Jatav Muslim Brotherhood,/ Where [the hell] did these Hindus come from anyway?
Jatav Muslim Bhai-bhai,/ Hindu kaum kahan se aye?

Instead they insisted that this slogan was raised by other parties fearing their defeat (*Amar Ujaalaa March* 27, 1962).

There was another instance of this during one of the elections for the deputy mayorship of Agra among the Municipal Corporators. The Republican Party put up the B.A. graduate of their own caste as their candidate; he was defeated and received only sixteen votes, all from his own party. This election was taken in an elective body, the Agra Municipal Corporation, where the Republican Party is second in number to the Congress, which itself does not have a clear majority. Subsequently, when the Republicans put up the Vaisya doctor as a candidate for this post, they were able to form a coalition with the Jan Sangh and some Independent members. As a result, the Republican Vaisya secured the votes to become Deputy Mayor.

In situations like this the Jatavs are activating the dominant status of citizen while the other castes are using the Jatav's salient, and in this case illegitimate, status of Camar or Untouchable. Such incidents also occur outside of the political arena. A teacher in a government college wrote to me: "Here in this college there is a lecturer in Sanskrit [a man of a merchant caste, or Vaisya] who refused to dine with me because of my being of a Scheduled Caste, though later he realized his mistake. He is still away from our mess." These are instances of "bridge actions," that is, actions in terms of either caste or democracy as convenience or belief dictates.

The Jatavs, too, do the same thing when they activate as dominant their status of Scheduled Caste. Such a status is dominant when it is a question of "protective discrimination." In such situations members of the upper castes attempt to activate citizenship as the salient status, and complain that there should be equality for all and special privileges for none.[40] In these cases, however, the government is not neutral between the two parties; it decides in favor of the Scheduled Caste status. Such action would be contradictory in a mature democracy, but is hardly so for a young one such as India.[41]

The general policies of the Republican Party are set out in its Election Manifesto (Republican Party of India n.d.). The major demands are concerned with improving the lot of the poor and "downtrodden" of India. A similar charter of ten demands was

[40] This is again the problem of which is primary, individual or social justice.
[41] For an opposing point of view and an extended discussion of this topic, see Donald Smith, *India as a Secular State* (1963: 292–332).

presented to Prime Minister Shastri recently deceased. These demands were:

1. The portrait of Baba Sahab, Dr. B. R. Ambedkar, "The Father of the Indian Constitution" must be given a place in the Central Hall of Parliament.
2. Let the land of the Nation go to the actual tiller of the land.
3. Idle and waste land must go to the Landless Labourers.
4. Adequate distribution of Food Grains and Control over Rising Prices.
5. Lot of Slum Dwellers be improved.
6. Full implementation of Minimum Wages Act, 1948.
7. Extension of all privileges guaranteed by the Constitution to such Scheduled Castes as embraced Buddhism.
8. Harassment of the Depressed Classes should cease forthwith.
9. Full justice be done under the Untouchability (Offenses) Act to them.
10. Reservation in the Services to Scheduled Castes and Scheduled Tribes be completed as soon as possible, not later than 1970 (Republican Party of India 1964: 1).

In addition to national resolutions such as these, the state branches of the Party generally make their own annual resolutions. The Uttar Pradesh Branch usually includes resolutions such as: that Urdu should be taught in the schools on a par with Hindi, if it cannot be made the official state language, and that taxes on shoe makers should be remitted. The first is a concession to the Muslims with whom the Uttar Pradesh Branch has close relations, especially in the city of Aligarh. The second reflects the prominence of the Camar group of castes, particularly those of Agra and Kanpur Cities, in the Uttar Pradesh Republican Party.

The ten resolutions listed above are instructive for a number of reasons. All of them are of immediate concern to the Agra Jatavs and come up frequently in daily conversation. The seventh resolution underlines again the close relationship between the Party and the Buddhist movement, just as the first resolution demonstrates the revered place Dr. Ambedkar has in the hearts of members of both the Party and the religion. The ten resolutions might be summed up as the "politics of Untouchability." They are the demands of a Party which feels it has too little of the already scarce economic, social, and political resources of the country. All of these demands are iterated and reiterated at rallies and meetings throughout the year. Politics among the Agra Jatavs is not just a matter of periodic elections; it is, on the contrary, a matter of rudimentary survival.

At public meetings of the Party few speeches are ever made without mentioning the condition of the Agra shoe makers, who were described in Chapter III. It was pointed out that many Jatav workers are in debt, either to the factory owners (even of their own caste), or to the hated factors in the market. The relationship of the Jatavs and the Punjabi factors is ideally a contract to cooperate for mutual advantage. In reality, however, the Jatavs feel that the relationship is one of ruthless exploitation.

To date, political representation has been able to channel the complaint of the exploited. Given such conditions, one might expect the Jatavs to be attracted to Communism. This seems to have been the case among the more thinking leaders, but the attraction died down because of the alternative form of action proposed by Ambedkar in the Buddist movement and the Republican Party. Ambedkar also explicitly rejected Communism. A Republican leader in Agra notes:

At one time people were attracted to Communism. But we could never be Communists now. Baba Sahab Ambedkar showed us the difference between Communism and Democracy. In Communism a man is not free to say as he wants, he is always watched by the state. We could never stand for this.

However, there seems to me to be something more basic to the rejection of Communism than the simple adoption of the alternative proposed by Ambedkar. Most Jatavs still see their exploitation as the result of the personal power and superiority of the Punjabis. They have not taken the next step (one they have already taken in regard to the caste system) of viewing the Punjabis' superiority as a partial result of the market system of economy.[42] Therefore, they have not sought an alternative system such as Communism, but rather they look only for new factors who will give them a fair deal, as did the now idealized Muslim factors of pre-independence India.

However, if conditions do not become more favorable or if they worsen, it is probable that more drastic solutions will be sought. Just as the former Sanskritizing movement of pre-independence days was rejected for the now more effective alternative of political participation, so too may the present movement be rejected for a more radical solution, if the Republican Party fails.[43] A number of leaders suggested this to me in statements such as the following:

(1) If at the time reservations end, the Republican Party has died, then all will turn to Communism. Where else is there any hope of fighting and an organization [to do it]?

(2) If the Party dies, then there is only Communism. Communism is where all people are equal.

These two statements are both germane and prophetic. They pellucidly demonstrate how conscious the Jatavs are of their goal of equality and how willing they are to turn to any organization which will fight on their behalf.

The tenth resolution in the Republican Election Manifesto is a subject of debate within the Party itself. This is the resolution which asks for an end to reserved jobs in the government services for the Scheduled Castes and Tribes. One group, the "moderates,"

---

[42] "A contract is ideally an agreement to cooperate for equal advantage. When it habitually produces unequal advantage, exploitation is suspected. The idea appeals strongly to the exploited, who quickly conclude that the power that is defeating them resides, not in the personal superiority of their oppressor, but in the unfair advantages he derives from the system. If the system renders contract a sham, the system must be changed" (Marshall 1965: 189).

[43] The lesson here for the present government is clear. It should support, rather than try to repress, the functional alternative to Communism with which the Jatavs are now experimenting. Their commitment to Ambedkar as the "Father of the Constitution" and the parliamentary form of democracy are in no way contrary to the Indian Republic as it now exists. The real danger to the present government lies not in supporting this movement; it lies, rather, in frustrating it.

feels that all reservations should be continued for some time. Another group, the "radicals," feels that, as a minimum, reserved elected posts such as M.L.A. and M.P. should be abandoned;[44] and as a maximum, all types of "protective discrimination" should be abandoned. The "radicals" maintain that because of the present system of reservations, Congress "yes men" are elected from the Scheduled Castes rather than men who will represent and fight for the "real" interests of the Scheduled Castes.[45] They also reason that all "protective discrimination" should be abandoned because doing so would force the Scheduled Castes to fight for their rights as citizens and to rely solely on their own efforts. The privileges that come with Scheduled Caste status, the radicals argue, only make Untouchables into "Uncle Toms" for the Congress Party. As far back as 1956, the president of the Uttar Pradesh Scheduled Castes Federation was reported to have said in a speech, "We don't want reservations; we want equality" (*Sainik* March 18, 1956). The policy of "protective discrimination" is, the radicals say, a Congress trick to divide and rule the Schedule Castes; if it is abandoned, then the Scheduled Castes will be forced to unite as a matter of political survival.

There is, moreover, a latent disfunction in the policy of "protective discrimination" for the Jatavs as a whole. The best educated members of the community are assured of posts in government service, and as government servants they are not allowed to engage in political activities. The caste is, therefore, deprived to a significant degree of those who are educationally best qualified

[44] The argument is not merely academic. The provisions of the protective discrimination policy were originally to last for ten years—from 1950 to 1960. In 1960, however, they were renewed for another ten years. Thus, what will be done in 1970 and what will happen thereafter are matters of vital concern to all Jatavs, as well as to all Scheduled Castes.

[45] This opinion is given support in an analysis of the Congress Party of Uttar Pradesh. The author, a political scientist, concludes: "The Scheduled Caste leaders who have been given Congress tickets in the reserved constituencies are non-militant and have no power in the local or state Congress organizations. The numerous organizations in Uttar Pradesh for the advancement of the Scheduled Castes and 'depressed classes' have been content to serve as agencies for the distribution of Congress patronage" (Brass 1965: 105).

to be its leaders. After my return from India, a college educated and strongly committed young man wrote to me: "The present government service is an obstacle in the service of the poor masses. So I shall be resigning from it in the near future." In effect, reservation of government jobs defeats one objective of the government's policy, because individual education and advancement do not necessarily redound to the benefit of the rest of the caste.

There is, furthermore, a difficulty inherent in the government's policy; it tends to perpetuate itself. The government has created conditions of unequal treatment for the Scheduled Castes in order that conditions of equal opportunity and treatment within Indian society might eventually come to be. Such preferential treatment, however, tends to become a vested interest, especially when an individual's immediate gain is placed against his caste's long range gain. Preferential treatment, then, perpetuates not only itself but also inter-caste (Dushkin 1957: *passim*)[46] and intra-caste rivalries. In the words of one informant, "The Congress Party's strategy with the reservations is like the British [policy of] 'divide and rule.' They want to create differences among us."

In view of this, then, the over-all political goal of the Republican Party and its present leaders is to defeat the Congress Party, because it is the party of the "Brahman boys" and the rich men. In short, the Congress Party is a negative reference group; it is perceived as an obstacle in the path of Jatav social, economic, and political progress. Thus, the aim of the Party is to unite all the Scheduled Castes against the Congress Party. This goal and its political meaning was artfully described by Ambedkar in one of his Agra appearances. His vivid metaphor remains on the tip of every Jatav tongue. Hindu society, he said, is like a wall built of four layers, the Untouchable being the bottom layer which supports the upper three levels. Remove the support of that bottom layer, and the whole structure of Hindu society will fall. This model of Hindu society contains much truth; it contains even

---

[46] The history and critical analysis of the whole subject of "protective discrimination" are treated at length in Dushkin (1957, and 1961a, b, c).

more truth as a model of Indian politics. If all the Scheduled Castes could unite into a single party, there is no doubt that they would be a significant threat to Congress Party rule.

Jatav demands are made known not only through formal presentations and public meetings but also through parades and acts of civil disobedience (*satyagrahas*). In 1963, a parade of Jatavs marched through the streets of Agra to the Divisional Commissioner's residence shouting as they went, *"roti, rozi, aur makan"* (food, work, and shelter). The parade presented to the Commissioner sixteen demands similar to the ten mentioned above. At the time, there was also much discussion about whether or not to go to Lucknow once again to conduct an act of civil disobedience for the release of Mr. B. P. Maurya, a Republican Member of Parliament who had been imprisoned under the Defense of India Rules (D.I.R.).[47] A plan was made to do so, but its execution was postponed because of half-hearted support and lack of organization by the Party leadership. Maurya was later released, and the project was abandoned.

The charge is often made by other castes that the Republican Party is a blatantly communalistic organization; that it emphasizes Scheduled Caste distinctiveness and uniqueness within the total population and is intent upon furthering only Scheduled Caste, or more accurately Jatav, interests. Paradoxically, such separatist policies have resulted in greater integration and articulation with the larger society. As Andre Beteille has perceptively pointed out, ". . . the measure of integration lies not so much in a passive acceptance of the *status quo* as in the adoption of a body of common political rules through which divergent interests are organized and articulated" (Beteille 1965b: 33).

The integrative function of the Party follows a double pattern. It unites castes horizontally on a regional and interregional scale. The Jatavs of Agra are in contact with Jatavs in the rest of Uttar Pradesh, as well as with other Scheduled Castes in eastern Uttar

[47] These were put into effect when a state of emergency was declared after the Chinese incursions into India.

Pradesh. Strong links with the Mahars of Maharashtra, who spread down to Mysore, also exist. It was hoped that a party unit would grow in Bengal under Jogendranath Mandal, former Law Minister of Pakistan. However, he was imprisoned under the D.I.R. before he could take any effective action. Identification through the Buddhist movement also links these castes into a single group with common purposes and a common identity.

Integration is not only taking place horizontally between castes but is also taking place vertically across caste boundaries. Politics and the whole governmental structure of patronage and development have placed various Jatavs into higher levels of state and national organization. The same is true of the schools, the administrative structure, and the market system of economy. A direct effect has been a gradual replacement of hereditary caste leaders (*chaudhuries*) by the "politicians" (*neta log*) and a disintegration of the urban caste council (*panchayat*) system. This last effect we consider in Chapter VI.

From a structural point of view, all these changes can be seen as the movement of the Jatavs into status positions formerly closed to them because of their controlling status of Camar. The Jatavs are now accorded *de facto* the right to occupy these statuses because of the new controlling status of citizen. Citizenship, not caste, is the prerequisite status for adding other statuses to status-sets. These new positions also allow pressure to be put on others to recognize citizenship not only as *a* controlling status, but also as *the* dominant status in actual interactions. In other words, *de facto* occupation of new statuses must also be accorded *de jure,* if only publicly so, by other castes; citizenship must take precedence, both behaviorally and normatively over other principles of recruitment and interaction. This pressure for structural consistency and openness in India is moving toward a system that is "cross cutting" and based on universalistic citizenship status rather than toward a system that is "mutually exclusive" and based on particularistic caste status.

Leadership

The rise of the Republican Party of Agra has brought forth a new type of leader called the "politicians" (*neta log,* literally the leader people). These men are distinct from the old "big men" who still exist as the wealthier entrepreneurs of the Jatav caste. The majority of "big men" today are Congressmen, and they form a third political bloc in the Jatav caste. This bloc I call the conservatives, in contrast to the two Republican blocs, the moderates and the radicals, of whom I have already spoken (see pages 106–07). On the whole, the conservatives do not engage in direct political action and therefore are not "politicians." They are a small elite group with little if any following among the Jatav masses.

The conservatives consider membership in the Congress Party a matter of survival. The Congress controls to some extent the financial resources and licensing offices upon which they depend. The aims and sympathies of the conservatives are with their caste, but their tactics are somewhat different from those of the "politicians," who are virtually all Republicans. The conservatives' main strategy can be summed up in the phrase, "Why bite the hand that feeds you?" [48]

Obviously, these men have defined their social situation differently from the Buddhist Jatavs. By identifying themselves as Jatav and Camar, they implicitly surrender themselves to the status of Untouchable. In this sense, then, they are not mobility-oriented. Their reference group of imitation is the other Scheduled Caste groups who follow the path of cooperation with and within the Congress Party. Therefore, in the field of politics their negative reference group is the Republican Party and other parties in opposition to the Congress because such parties block their paths of action. As one of them testified to me:

[48] The parallels in the development of these two types of leadership with the development of the leadership among the Eta of Japan are so striking as to seem more than coincidental. The development of Eta leadership groups is treated in a paper by Cornell (1963) entitled "From Caste Patron to Entrepreneur and Political Ideologue: Transformation of Nineteenth and Twentieth Century Outcaste Leadership Elites."

Republican Party parade with a banner celebrating Dr. Ambedkar and his deeds.

We [the conservatives] have the organization and the money but we cannot accomplish anything until we have the backing of the people [the Jatav masses]. This we intend to get so that we can present our demands forcefully to Congress, who alone can help us now. If we have the people behind us, Congress will meet our demands.

The political position and strategy of these men was made painfully clear by one of their leaders:

I left the Republican Party almost three years ago. What is necessary first is education and businessmen who have some money. When we are a minority community without any power, then it is not for us to openly criticize those in power and to revile them. Therefore, we should cooperate with them as much as we can now and get as much as we can now. We should take advantage of the provisions for education and business which they offer to us now, before it is too late.

If I carried on as some of the members of the Republican Party do, where would I be now, as a businessman? I have made much progress by not following its way. The Hindus are in power and we are too weak to stand up to them. Only when we have educated men and powerful businessmen of our own can we stand before them as equals.

The above statement clearly demonstrates that the conservatives have acquiesced to social inequality by accepting the status of Scheduled Caste with its implicit status of Untouchable. Unlike most other Jatavs, they are subject to a conflict within their status-set. Compared to other Jatavs, they are economically rich and they feel that their economic status is dependent to a large extent upon membership in the Congress Party and the largesse of upper castes within it. They feel that their economic patronage will be lost if they do not accept in the political field the salient Uncle Tom-like Scheduled Caste status accorded them by the upper castes that control Congress. They are, in a sense, urbanized and modernized clients (*purjans*) of upper caste patrons (*jajmans*).[49] Other Jatavs have a poor but politically independent economic status and have nothing to lose if they aggressively assert their dominant status of citizen in the political and social fields. However, they too have

---

[49] A *jajman* is a patron and a *purjan* is his client in the *jajmani* system.

something to gain when they accept their Scheduled Caste status to qualify for welfare patronage.

Because the government has limited economic patronage to hand out, the plurality of Jatavs who are not recipients of it can act with political independence. Economic patronage, which the conservatives receive, is subject to the particularistic selective criteria of political contacts and differs from welfare patronage which is largely subject to the universalistic criteria of civil service exams. Moreover, since the quota of government jobs reserved for Scheduled Castes is still largely unfilled, this cannot be considered a scarce commodity. Universalistic criteria also largely determine the allocation of educational benefits given to the Scheduled Castes.

A further point must be emphasized here. The conservatives do not differ from the plurality of Jatavs in the opinion that their social status ought to be higher than it is. The difference, rather, is in behavior; they openly cooperate with and accept upper castes and the Congress, while other Jatavs are in conflict with and reject upper castes and the Congress. As Berreman (1966: 288) has noted, "Consensus is not crucial to successful interaction but . . . behavioral articulation is essential." The conservatives place the achievement and maintenance of a higher economic position before the achievement and maintenance of a socially equal position.

The "politicians" on the other hand, although they all have some independent means of support, are primarily politicians. They are no longer just "big men," as were the leaders of pre-independence days. These men deal in patronage, power, and influence which are based upon the structural positions they now occupy and the functions they now perform.

Most of the "politicians" (party leaders and officers) occupy elected governmental offices. It is these elected men who form today the *de facto* Jatav elite. Included in this group are the municipal corporators and the members of the State Legislative Assembly. They hold these statuses on achieved not ascribed criteria. As members of these assemblies, they serve on various committees and are thereby in positions of structural observability from which they can monitor the actions and decisions of other members of these committees. While the committees have little formal power or authority, they do have a considerable amount of informal power. Committee members in the Agra Corporation can reduce assessments on house taxes and decide on development projects, for example. The importance of all this is twofold. First of all, structural observability is translated into communication with other members of the caste in such a way that all Jatavs are aware of what decisions are being made, as well as of the substance and effect of these decisions upon them. Second, the Jatavs, as citizens, occupy these positions on a level of equality with members of other castes. In these statuses, therefore, relations are manifestly co-ordinate, not subordinate. Thus, other castes are forced to bargain and to work with Jatavs in ways not possible in the traditional caste system.

One example of this is the pact, already mentioned, among the Republicans, the Jan Sanghis, and some Independents to elect a mayor and a deputy mayor of the Agra City Municipal Corporation. Through this pact an Independent of Vaisya, the merchant group of castes, became the mayor, and a Republican became deputy mayor, although he is not a Jatav. This coalition succeeded in defeating the Congress candidates.[50]

It is interesting to note how in the 1962 elections the Congress Party responded to this coalition by playing caste politics. It activated caste as a salient status and soft-pedaled the relevant and normatively dominant status of citizenship. The Congress Party Jatavs, through their Scheduled Castes organization, the Depressed Classes League (*Dalit Varg Sangh*), issued a circular which read as follows:

We humbly request all the downtrodden people and especially the Jatavs to seriously consider, in casting their vote in the General Elections on March 12th, with which party candidates their good lies.

[50] A history of the many factions and alliances in the Agra Corporation is treated in Rosenthal (1966).

Some persons of the Scheduled Castes, who are in positions of power [the "politicians"] and who call themselves policy makers and leaders of the Jatavs, want to sell their Jatav brothers for their own selfish ends. It is because of this that they have entered into a contract with the Jan Sangh, those communal, narrow-minded, and primitive sloganeers of the Brahmanical Hindu State. These leaders have not even consulted the voters, their Jatav brothers.

We cannot understand how there can be a union of straw with fire. These devotees of Brahmanism, these priests, these primitive Jan Sanghis and their kind, have continually been inflicting all sorts of tyranny, injustice, insults and disrespect upon us, according to the Laws of Manu Smrti (Dalit Varg Sangh n.d.).

This is truly an amazing document, for it demonstrates that beneath the lid of Congress membership boil the same emotional attitudes and resentments of their openly defiant caste mates. It echoes with the true bark of those who, in the eyes of most Jatavs, take a tail-between-the-legs attitude by joining the Congress Party. Indeed, this document reads more like a polemic from the embittered pen of Dr. Ambedkar than a political pitch from the smoky back room of Congress conservatives.

It is important to note once again that it is the Municipal Corporators and the M.L.A.'s, not the officers of the Party, who hold effective power. Conflicts have arisen between the elected representatives (called the *dal,* that is, faction or bloc) and Republican Party officers. In one such instance in 1964, pressure was put on the Party president to force the Republican deputy mayor to resign in accordance with the pact the Party had made with other parties in opposition to the Congress. The Party's president, a very upright man, felt the deputy mayor should resign lest the Republicans get a bad name and never again be trusted by other parties. The deputy mayor and his faction considered this matter in a more pragmatic and politically expedient frame of reference. They elbowed in their own interpretation by finding a legal loophole in the way the pact was written and enacted by the other parties involved. The letter of the law was more important than the spirit. The deputy mayor was able to get the Republican M.P.

from Aligarh, who is also vice president of the state Party, to come and give a speech in Agra. In the course of his speech, the M.P. came out strongly in support of the deputy mayor against the Agra Party president and his faction. The M.P. is an inspired orator and his charisma weighs heavily with the Jatavs. Thus, a word from him signaled the virtual defeat of the opposing faction. Furthermore, those in offices of political power have access to real patronage and, thus, are able to do something concrete for their constituents. The Party officers, unless they also hold some elected post, have no such power and are in a far weaker position to gain support for their views.[51]

The position of the Republican Vaisya, who was elected deputy mayor, is intriguing. In order to get himself elected he needed Jatav votes from his electoral ward. These he got by joining the Republican Party and contributing to it financially from his not inconsiderable wealth.[52] The Jatavs in turn use his upper caste status to assert that they are not a one-caste communal party. Most importantly, his higher caste status, wealth, and influential connections allow him structural observability over informal situations where Untouchable caste status is salient and Jatavs may not tread. "Indeed, some upper-caste corporation members rather relish the possibility that they are contributing to the disruption of the Republican Party by strengthening the position of . . . [Agarwal] within it" (Rosenthal 1966: 377). Thus, he is a valuable liason with the informal structures of power and influence where upper caste status is his pass-key, so to speak. Finally, he is a man of experience and wealth, and as such he is the virtual leader of the "politicians" in the Republican Party of Agra. In effect, he is

[51] The same factor has been noted in a voting study of Farrukhabad district, Uttar Pradesh, for the Congress Party. ". . . every visitor to the Congress office asks for the Congress M.L.A. and not for the office-bearers of the Congress Party. This indicates that the Congress as an organization does not have much significance in the public eye; it is rather the Congress M.L.A.—a symbol of the ruling party and a dispenser of petty favors" (Roy 1965: 895).

[52] He is also a doctor and a real estate speculator and developer. During the election there was a rumor raised that his mother was really a secret Camar. Just how many votes this rumor actually won or lost will probably never be known.

the "outside leadership" which Weiner has so aptly described as characteristic of many secondary organizations in India.[53]

Generally by reason of their income, education, family background, land ownership, or caste they are of higher social status than the groups they lead. Of the various hypotheses, status is perhaps the most persuasive and most adequately explains the general phenomenon. Although India has a hierarchical social system in which subservience to authority and acceptance of one's role, no matter how onerous, are important values, people have not been reluctant to protest through political organization. But in organizing politically, they turn to those of higher status for their leadership (or at least are organized by them) (Weiner 1963: 99).

The activities of the "politicians" are three: goal achievement, interest articulation, and better organization. By goal achievement is meant that the "politicians" have made substantial gains for their party and caste through political participation and representation. The first and most important gain relates to the channeling of some development funds to their own electoral wards. The municipal corporators have the right to decide which neighborhood will receive these development funds and in making these decisions they rarely forget their own caste neighborhoods. Thus, in many Jatav neighborhoods one can now find electric street lamps, brick-paved alleyways, and additional water outlets. Second, a picture of Dr. Ambedkar alongside other national and local notables now hangs in the meeting hall of the municipal corporation. This picture would not have been hung there unless the Republicans had fought for it. Also, there are resolutions pending to erect a statue of Ambedkar at Tikonia Bazar in the center of the city and to rename a road Ambedkar Boulevard (*Marg*). The public display of these symbols is also a forced public recognition not only of the Untouchable Ambedkar but also of Jatav political muscle. I say that this is forced public recognition because, *vis-à-vis* upper castes, Ambedkar's and the Jatavs' salient

[53] Marshall (1965: 98) describes an almost identical situation for nineteenth century England. This would suggest that the sociological variables involved here are not arbitrary.

status is Untouchable. One of the local newspapers reports upper caste reaction to these Republican demands:

People are objecting to this tendency, which is widespread among the Republican members of the Corporation. This tendency gives the appearance that Baba Sahab Ambedkar's name may be put on every street, library, park, road and school and, wherever there is room for a statue to be placed, there Baba Sahab ought to be placed. In this way they want to flatter the minds of all people with Dr. Ambedkar. . . . But here the other parties of the Corporation are putting a stop to the continuance of these trends because of Shudra political self-interest and are containing them (*Amar Ujaalaa* August 10, 1961).

Third, Jatavs consider it no mean feather in their cap that members of their caste have been elected to offices in the local power structure. When I playfully provoked one informant with the charge that the Jatav corporators are scoundrels, he retorted, "No matter what they do we'd elect them again, just to show the Hindus that we can do it." Note once again how this informant spontaneously opposed two groups: we Jatavs and those Hindus; we Jatavs are *not* Hindus.

Interest articulation, the second activity of the "politicians," involves making the Republican Party's demands and views known in the local, state, and national legislatures as well as before government administrators. Those in statuses of some power and those with access to public forums make grievances known and intercede for their Jatav brothers and other members of their constituencies in the courts and with the administration. A typical corporator told me: "I was elected to bring justice to my ward. Before this there were no paved streets and the Congress leader exploited but didn't help us. We had trouble with the police. For this reason I was elected, and these problems I've been able to solve to some extent."

Interest articulation also includes the airing of Jatav complaints, desires, and wants at the many public meetings held during the year. Two functions are involved. First, these complaints openly display the Jatav political stance to members of the government and to other castes. Second, these public declarations serve to reinforce Jatav solidarity on the issues discussed, and they bring new information and ideas to the group as a whole. Below are some verbatim excerpts from my notes on 1963–64 speeches. They illustrate not only the substance of Jatav protest but also the latent function of current politics in integrating the Jatav community into the nation, the state, and even into the world community.

Speech I
. . . opened with an expression of grief over Kennedy's death. He gave much praise to him as a fighter for peace and for the rights of all men, not just for America. He said India owed much to him. Then he spoke of education. He said *rupees* 80 *crores* was needed to give basic primary education through the country, but that Congress and the upper castes really did not want this. They spent *rupees* 42 *crores* to open a new air-conditioned medical college in New Delhi, but not on basic primary education. . . . He said that Hindi should be the language of the country but that upper castes really used it to their advantage because they send their children to learn English and then this [not Hindi] becomes the criterion to get service in the government.

. . . As for agriculture, he said that the government foolishly sends experts for a joy ride to the United States to learn about farming where one man farms 10,000 *bighas*[54] by machine, while in India a man owns 10, 20, or 50 *bighas*. People ought to be sent to Japan, where small-scale farming is carried on that could benefit India.

Speech II
If we Scheduled Castes are given the facilities today that the Negroes in America now have, we could call off our movement. It is false when Congress tells us we are better off than they. In fact, they are better off than we; they are strong and united and have some money. We have none of these. . . . What kind of democracy is this when I am put in prison for saying the same things the Prime Minister and Jayaprakash Narayan say?

Other, and perhaps more important, channels of communication in addition to the formal meetings of the Party exist. Most important of these are the shoe market places, the shoe factories, the Party newspapers, and assorted pamphlets which are passed

---

[54] A *bigha* is equal to one-fifth of an acre.

from hand to hand. There are two Party newspapers, the *Republican News* (*Sandesh*) from the neighborhood of Aligarh City, and the *Stars of the Soil* (*Zamin Ke Tare*) from Aligarh City itself. The former is published irregularly and the latter comes out bimonthly. In Agra, too, a series of newspapers have been published since 1956, but all have been shortlived.

The final activity of the "politicians," that of better organization, consists of bringing together in united effort both Party and caste when specific actions require it. Thus, the various acts of civil disobedience, protests (*andolans*), and meetings throughout the year depend upon the decisions and organizational skills of the "politicians" for their success. A striking example of this was the First All-India Buddhist Conference (Bodh Sammelan) at Agra in 1963 (see Chapter V). The Party supplied the leadership and organizational skills to make this meeting a success after Buddhist leadership proved unable to do so.

Political Liabilities

Thus far the analysis has concentrated primarily on the external relations of the caste and the Republican Party to other castes and parties. On the whole, I have presented the movement as though it were completely united and organized. There are, however, four major disabilities from which the Party suffers and which sap away its very life substance. These are internal factionalism, poor leadership, lack of money, and communalism.

From its inception in 1958, the Republican Party of Agra has been racked, but not rent, by internal faction and discontent. The original quarrel arose when the Scheduled Castes Federation (S.C.F.) was dissolved and the Republican Party of Agra (R.P.A.) was organized in its place. One group, the moderates, held that the former officers of the S.C.F. should continue to hold the same offices in the new R.P.A. Another group, the radicals,[55] claimed they had elected new officers for the Party. The radicals were

---

[55] I have already identified a third political bloc within the caste. This is the conservatives, who are composed, for the most part, of "big men" who have joined the Congress Party.

abetted in their aims by the Secretary of the Party for the state of
Uttar Pradesh. From that day until the 1959 elections, both groups
tried to gain recognition and legitimacy from various state and
national officers in the high command of the Party. However, in
the interests of winning the 1959 elections to the Agra Municipal
Corporation, the total command of the Agra Party was entrusted
to Mr. Man Singh, a man of wide experience and great organizational ability. Much of Jatav success in that election is justly
credited to him.

After the 1959 elections the feud broke out again. Mr. Man
Singh himself was accused of taking and using personally a sum
of money donated by Dr. Agrawal, the merchant caste member of
the Party. This money, it was said, was meant for the Party's
election campaign expenses. Charges of corruption were leveled
against Mr. Man Singh. He resigned (*Amar Ujaalaa* March 30,
1961), and was later expelled from the Party for thirteen years
(*Sainik* July 5, 1961). In expelling him, the Party lost its most
capable Jatav leader. A new president was then chosen by lottery.

After the 1962 elections for M.L.A. were over, the new president resigned, but not before appointing candidates from among
the radicals to fill party offices. These candidates were from various
castes; there were two Jatavs, one boatman (*Mallah*), one sweeper
(*Valmiki*), and one Christian. The strategy of the radicals was to
convert the Party from a single-caste party into a multi-caste party
in order to win a broad based support for the Republican Party
in the next Municipal Corporation elections. The bloc of moderates, however, rejected this slate and put up an all-Jatav one of
their own. The radicals allege that the moderates' action was
motivated by the desire to keep Party power in Jatav hands and
in so doing they destroyed any hope of a mass appeal for the
Party.[56]

Into the midst of this internecine conflict stepped the Secretary

[56] Morris-Jones takes note of Myron Weiner's explanation of such a phenomenon,
and then gives his own more general interpretation. I believe Morris-Jones's explanation is the more correct. He says: "Myron Weiner in his *Party Politics in India*
offered the explanation that in many cases party had become a substitute for *jati*

of the Party for the state of Uttar Pradesh. He immediately dissolved both groups and set up a committee to elect a new president on an all-city basis. 1,553 members from every Jatav neighborhood in the city were enrolled in the Party. These members elected neighborhood chairmen who, on October 5, 1963, elected the present president, a man of sufficient honesty, integrity, and age to be respected by all. His election was hailed as the first real election of a Republican Party president in a broad based democratic way.

The fight between the Party officers and an opposing faction has already been mentioned (pages 116–17); therefore, only one rump group remains to be discussed. After his expulsion from the Party, Mr. Man Singh joined the Congress and worked for it during the 1962 election. He thereby ruined any chance for forgiveness; his act was considered one of a traitor, and he is despised for it to this day. When Man Singh, at a public meeting, accused the Republican Party of following the path of corruption and of having a secret relationship with Pakistan, the hatred for him and his sacrilege of joining the Congress Party violently erupted. A local newspaper describes the event thus:

Present in the meeting were some people wearing blue caps,[57] and these are reported to have been connected with the Republican Party. The complaint of the Congressman is that during the meeting these people shouted slogans and, throwing bricks into the meeting, they obstructed it. Two or three people were injured by the bricks. . . . Some police were present on the occasion and later the riot squad came (*Amar Ujaalaa* January 15, 1962).

Man Singh got nowhere with the Congress and subsequently

---

and that what members demanded above all from their political group was the snug and reassuring coherence of a unit in which there are no strangers or outsiders. A more general, though not incompatible explanation would be that even in the sophisticated world of urban party politics men have not so far left behind the atmosphere of traditional politics that individuals can be taken as persons; on the contrary, frankness is killed by suspicion and every disagreement becomes a situation of intolerable distrust" (Morris-Jones 1964: 64).

[57] The blue caps referred to in the account belong to Republican Party members. Blue is the Party's color, and Republicans wear blue caps just as Congressmen wear white ones.

left it. Together with one of the Jatav Municipal Corporators, he is now trying to form a new group called the Ambedkar Forward Bloc. His bad name and the Corporator's dubious reputation give the movement little chance of gaining a following.

These power struggles within the Party are very informative. In the first place, they point out that the Party itself is a structure within which and through which an individual can become socially mobile. Status in the Party confers higher rank within the caste, and election to a legislative post through the Party puts a man into a high rank leadership position within the total society. In the second place, they underline the ambivalence of members of the caste between the latent caste norms of the moderates, who would make the Party a caste "interest group," and the manifest class norms of the radicals, who would make the Party a true party of the minority groups and the poorer classes.

The second liability which threatens the Party is the poor quality and qualifications of its leadership. The Party has few men with real, proven organizational ability. In this sense, the expulsion of Mr. Man Singh was a great loss. Dr. Agrawal, the merchant caste member of the Party, has the ability, time, and the money to perform these organizational tasks, yet he does so only when it is absolutely necessary. Like other Party members, he is not noted for works of supererogation on behalf of the Party or of the Jatav caste as a whole.

The death of Ambedkar, before the Party had taken its first halting steps, left it to find its own leaders. None of the present leaders have the charisma with which Ambedkar so entranced the masses. The Republican Member of Parliament from nearby Aligarh City, B. P. Maurya, is now the *de facto* leader of the Party in western Uttar Pradesh. He is known for his fiery, witty, and bitterly anti-Congress speeches, and his appearance on an Agra stage always draws a large and sympathetic audience. However, he too is increasingly subject to the criticism now leveled at all the leaders, that they are corrupt and open to bribes. Only the present president of the Agra Party escapes such accusations. Unfortunately, he is too mild-mannered and sincere a man to be a really effective leader.

Part of the problem of poor leadership is lack of education. Only one of the Jatav Republican M.L.A.'s, who is also a member of the Municipal Corporation of Agra, is a college graduate.[58] The average number of years of formal schooling for ten of the Jatav Municipal Corporators is six, but the range of variation is great (17, 12, 9, 9, 8, 8, 4, 1, 0, 0). Lack of education is held against them by members of other castes, with the result that the Jatav leaders have been stereotyped as ignorant and not to be taken seriously. This stereotype exists despite the fact that a number of upper caste Corporators are as uneducated as the Jatavs.[59]

Leadership also suffers, because there are so few Republican M.L.A.'s in the State Legislature that they have the power to do little more than give token speeches of protest. The real source of patronage and legislation lies with the Congress Party.

The third liability of the Republican Party in Agra is its empty coffers or lack of money. The Party is, on the whole, a party of the poverty-stricken. There are no funds available in Agra to maintain an office or to keep a full or even a part-time staff; nor are there ample funds for conducting election campaigns. Without an office and a staff, there is no visible, formal locus of continuity for the Agra branch of the Party. Also, the Party's sources of patronage are very limited. In such a situation its appeal is restricted to caste and party loyalty. Furthermore, many of the leaders are by no means wealthy, although they generally have some independent means of support. This may explain the allegations that they are always open to bribery. My own evidence tends to support this contention. Thus, confidence in Party leaders has been weakened by corruption, and the rank and file have been given an example to follow suit at election times.

Finally, the greatest drawback of the Party is its membership,

[58] Another Jatav M.L.A., who is also a college graduate (he holds an M.A.), is a Congressman, but he is not an effective leader.
[59] For a detailed and penetrating analysis of the background of all members of the Municipal Corporation, see Rosenthal (n.d.).

drawn by and large from one caste only, although this varies by regions. For example, the Republicans of Maharashtra are almost all Mahars (an Untouchable caste), and the Republicans of western Uttar Pradesh are mostly Jatavs. This instills in the hearts of other Scheduled Castes in Agra a fear of Jatav domination in the Republican Party, just as many Jatavs fear upper caste domination in the Congress Party. The association of the Party with the Buddhist Movement has also alienated other castes. A prominent member of the Backward Classes told me:

Most of all they want those who enter the Party to convert to their religion. That is bad. Dr. Ambedkar gave them some poisonous injections to wake them up, but they have let it run wild. It is out of control. They are alienating all those who sympathize with them because of their actions and hate for others. They are a one-caste party and look down on those they consider below them.

While this statement is exaggerated, it does give some idea of how others interpret the Republican Party. What is really at stake is control of the Party. Potential members from other castes demand a share of the Party leadership, but the moderates now in control are reluctant to grant this to them. Thus, while internal homogeneity is eu-functional for the caste, it is dysfunctional for the Party, which must broaden its appeal to heterogenous castes in order to gain effective political power.

Summary

The Jatavs, then, have changed from mobility through Sanskritization to mobility through active and separate political participation. In modern, independent India political participation is a functional alternative to Sanskritization. The adoption of this new alternative is related to changes in the total structure of Indian society. Previously in a caste society based on caste rank, many aspects of the Jatavs' social condition such as total population, residential segregation, and a literate, enlightened leadership remained only adaptive potentials. Now in a parliamentary democracy, they have become actually adaptive because of the new statuses of citizen and voter. When the Jatavs use their voter status, the collective power of their numbers, their *en bloc* caste wards, and their enlightened leadership, they constitute a significant political entity. These political assets have been institutionalized into the Republican Party of Agra.

Jatav response has also been conditioned by the way in which most of them have defined and redefined their social situation in terms of reference groups of imitation, identification, and negation. Much of this reinterpretation was due to their charismatic leader, Dr. B. R. Ambedkar.

It is clear that what is going on in Agra between the Jatavs and the upper castes is an instance of what Geertz has called a "civil politics of primordial compromise" (Geertz 1963: 157). In such a politics we find that, "Alongside of, and interacting with, the usual politics of party and parliament, cabinet, and bureaucracy, or monarch and army, there exists nearly everywhere, a sort of parapolitics of clashing public identities and quickening ethnocratic aspirations" (Geertz 1963: 124). We have seen this in the contradictions that arise over which status (caste or citizenship) shall be dominant in actual interactions and whether caste or citizenship will be accepted as a controlling status for adding statuses to status-sets.

The importance of the *varnas* as categories open to achievement even within the traditional caste system has been noted. They represent "not so much a classification of caste ranks as a set of alternative interactional strategies . . . [by which] a caste may rise or fall many steps of rank . . ." (Marriott 1959: 68). The dismissal of these categories as sociological anachronisms is unwarranted.

Finally, the ways in which religion, economic position, and politics mutually influence one another in Jatav life has been described. Buddhism is an esoteric system providing a set of fundamental values and a model of ideal society. To make these a reality the Jatavs have taken to political action. It is, however, their low and exploited economic position in the market system which provides the bedrock issue for political nostrums on the here and now or, as we have called them, the exoteric issues.

CHAPTER V

Buddhism and the Bodhisatva, Ambedkar

Twenty-five hundred years ago in India Buddha Shakyamuni received enlightenment under the Bodhi Tree. From that day until this the way of the Buddha has continued to challenge men's minds and motivate them to compassionate action. Yet, the soil that gave birth to and first nourished the eight-fold path could not sustain it for long. Buddhism, for all practical purposes, had died out in India by the 10th century, but not before it had sent out roots which found more nourishing soil in other Asian lands.

Within the last decade, however, there has been a resurgence of this ancient faith in India. Neo-Buddhism, as it is called, is found primarily among certain castes of Untouchables, among whom are the Jatavs. While the phenomenon of individual conversion is interesting in itself, conversion is even more interesting when it becomes a mass movement of a particular social group, as it did among the Jatavs of Agra. Intriguing questions immediately arise: Why was Buddhism, a faith long dead in India, chosen before all others such as Islam, which preaches the equality of all men before God, or Christianity which could bring material benefits through its missions and schools? How deep, widespread, and organized is this conversion and what is its future? What relation does Neo-Buddhism in Agra have to traditional Hinduism? Is it really a radical break with the past?

The Bodhisatva, Ambedkar

The conversion of the Agra Jatavs can be traced to the influence of one man, Dr. B. R. Ambedkar. Ambedkar not only led the conversion to Buddhism, but also wrote its bible, *The Buddha and his Dhamma (Ambedkar* 1957). He has become the culture hero of the Jatavs, and the attention he receives borders on worship. This is at first surprising because Ambedkar was not a Jatav from Uttar Pradesh; he was, on the contrary, a Mahar from the state of Maharashtra. These two facts, different caste and different regional affiliation, might ordinarily disqualify a man for leadership in India where such primordial loyalties run deep and sharply separate one group from another. Much of the impetus to the Jatavs' unusual selection of Ambedkar as their leader lies in the similarities of their life histories with his. Ambedkar's life epitomizes the painful realities, knotty problems, and radical solutions which have been part of their own lives; it was, thus, easy to transcend narrower loyalties of caste and region by identifying with him.

Bhimrao Ramji Ambedkar[1] was born at Mhow in central India on April 14, 1891. His father was a retired army man and headmaster in a military school. The family belonged to the Mahars, who were and are the largest Untouchable caste in the state of Maharashtra. Mahars traditionally were village servants and menials.

Ambedkar's early education was completed in Bombay, where he received the B.A. degree from Elphinstone College. His earlier school years were not without experiences which branded the

[1] Unless otherwise noted this section of the chapter on Ambedkar's life is based upon his biography by Dhananjay Keer (1962). Where the material differs from Keer, it is based upon Zelliot (1966). Miss Zelliot is now writing a definitive biography of Ambedkar and has graciously offered these corrections.

stigma of being an Untouchable on his mind. He was made to sit in a corner of the class separate from other students. For fear of pollution his teachers would not touch him or even ask for a recitation. And, like an armless beggar, water was poured into his mouth from above, lest his lips touch and pollute the container (Keer 1962: 14–15).

In 1913, the enlightened Gaekwar of Baroda sent Ambedkar to the United States for further studies. He enrolled in Columbia University in New York and received both the M.A. and the Ph.D. degrees in economics. At Columbia, "He could read, he could write, he could walk, he could bathe, and he could rest with a status of equality" (Keer 1962: 27). This experience no doubt colored his later life and his fight for the rights of the Untouchables.

Upon his return to India he was appointed military secretary to the Gaekwar of Baroda, for whom he had agreed to work in return for support of his studies. Yet the experiences he had to undergo in the Gaekwar's service forced him to quit in short order. Once again, as in his early school years, Ambedkar, a Ph.D. who had lived as an equal to others while abroad, was treated like a leper. Notes and files were flung at his desk lest the clerk, poor in education and lacking in foreign travel but high in caste status compared to Ambedkar, come into contact with the defiling doctor. Even drinking water, though available to his subordinates, was not available to him in the office (Keer 1962: 34).

I mention these details of Ambedkar's early life because they have become part of the halo of legend which surrounds him in Agra today. The Jatavs feel that Ambedkar's experiences and their own are identical. Because of this they follow him and adopt his methods and ideas for overcoming their own problems. Legends, like ideas, sometimes have consequences.

In 1920 Ambedkar returned to London where, in 1923, he received the D.Sc. degree in economics and was called to the bar. From this time on, Ambedkar's reputation and fame among the Depressed Classes, especially those of Maharashtra, began to grow.

In 1926 his nomination to the Bombay Legislative Council put him in a clear leadership position.

In 1929 the British government held a series of three Round Table Conferences to discuss the problems of a new constitution and a new system of government for India. Ambedkar was nominated as a representative of the Depressed Classes and attended all three conferences. The Congress and Gandhi, however, boycotted the first Conference in 1930, but agreed to attend the second in 1931. It was in the second Conference that the bitter and life-long enmity between Gandhi and Ambedkar first erupted.

At this Conference Ambedkar changed his political demand from one of joint elections with reserved seats in the legislatures to one of separate elections for the Depressed Classes. Gandhi, on the other hand, was inalterably opposed to such a policy. The Mahatma felt that separate elections—one for the Untouchables, another for the caste Hindus—would split the Hindu religious community as Luther's propositions had split the Christian community. For Gandhi the whole question of Untouchability and the Untouchables was a religious and moral one, and had to be kept out of politics. Gandhi accepted the *varanashram,* or theory of the four *varnas* of Hinduism, and conceived of the removal of Untouchability as the absorption of the Untouchables into the fourth, or Shudra, *varna.* He asked that this be done by the repentance and self-purification of the upper castes for their unjust and unequal treatment of the Untouchables. The well-to-do upper castes were to be the trustees of the poverty stricken lower castes.

Gandhi's philosophy and his program for the removal of Untouchability were understandably an anathema to Ambedkar. For him the question of political safeguards for the Untouchables was the crucial matter. He felt that the upper castes had made and kept the Untouchables a community separate and unequal from other castes and that in reality they were not part of the Hindu community. Unlike Gandhi, he saw no possibility that the Untouchables would either be assimilated into the Hindu community or be granted rights equal to those of upper caste Hindus. Thus, he concluded that separate elections alone would protect the Untouchables from upper caste dominance and disenfranchisement (Zelliot 1966: 198–99).

A result of this conflict was that from the time of the Round Table Conferences until Gandhi's death, Ambedkar considered the Mahatma an enemy, not a liberator, of the Untouchables (Zelliot 1966: 199). He once wrote:

To the Untouchables, Hinduism is a veritable chamber of horrors. The sanctity and infallibility of the Vedas, Smritis and Shastras, the iron law of caste, the heartless law of Karma and the senseless law of status by birth are to the Untouchables veritable instruments of torture which Hinduism has forged against the Untouchables. These very instruments which have mutilated, blasted and blighted the life of the Untouchables are to be found intact and untarnished in the bosom of Gandhism (Ambedkar 1946: 308).

Telegrams supporting Ambedkar's position came to London from Depressed Class associations throughout India. However, since no solution to the communal problem, including that of the Depressed Classes, was reached, the British premier asked all parties concerned to sign a statement saying that they would accept his decision in this matter. Gandhi signed but Ambedkar did not; he believed that the rectitude of his demands could not but be accepted. Ambedkar returned to India and to national leadership of the Depressed Classes. But to those who followed Gandhi, including the Congress, he returned as a traitor. Yet many felt that had Gandhi been willing to negotiate with Ambedkar, he could not only have settled the question earlier, but also could have won the doctor's support in negotiations with the British and the Muslims (Dushkin 1957: 64).

On August 14, 1932, the British announced their decision in what is known as the Communal Award. According to this plan, the Depressed Classes were to have a number of special seats assigned to them for a period of twenty years. Only Depressed Classes could vote for these seats, although they were also entitled to vote for the general seats. This was a victory for Ambedkar,

Dr. Ambedkar arriving at Poona in 1952 to give a speech on parliamentary democracy.

who held the view that unless separate elections were granted, the achievement of adequate representation in the Legislature by the Depressed Classes would be most unlikely.

Gandhi rejected the Award and began what is known as his "Epic Fast" against it. By so doing Gandhi, who claimed to be the true representative of the Depressed Classes, put his life in the hands of the man whose actual leadership of those Classes he denied. Ambedkar wryly noted: "The surprising fact is that my position as the leader of the Untouchables of India was not only not questioned by Congressmen but it was accepted as a fact. All eyes naturally turned to me as the man of the moment, or rather, as the villain of the piece" (Ambedkar 1946: 88). With Gandhi near death, a solution was finally reached in what is known as the Poona Pact. According to this Pact, the Depressed Classes were to be allowed a separate electorate in the primaries but they were to vote in joint electorates in the election itself. However, 18 per cent of the seats in the Central and Provincial Legislatures were to be reserved for Depressed Class members (see Ambedkar 1946: 88–89). The Congress was, thereupon, morally and politically committed to this Pact, although it neither accepted nor rejected it publicly (Dushkin 1957: 76).

Ambedkar's major concern during the late thirties and forties was securing political rights and representation for the Scheduled Castes. In 1941 he was appointed to the Defense Advisory Committee of the Viceroy of India. Then, in 1942, he was appointed as Labour Member in the Executive Council of the Government of India. This was the first time an Untouchable had reached one of the highest posts in the government of India.

In 1942, too, Ambedkar decided to consolidate and nationalize his leadership of the Scheduled Castes. Thus, in July of that year he founded the All-India Scheduled Castes Federation at an All-India Depressed Classes Conference in Nagpur. This was ". . . a direct appeal for the Scheduled Castes to win power through unity" (Zelliot 1966: 201). The basis of the Federation's aims were Ambedkar's fundamental and life-long tenets of political action.

The Federation was separatist in its claim that the Scheduled Castes were a separate and distinct group in the Indian population. It was also political in its efforts to gain economic and social rights for the Scheduled Castes.

The victory of the Congress party and the utter defeat of the S.C.F. in the election of 1946 ". . . virtually eliminated the threat of political Scheduled Caste separatism that had previously influenced Congress policy" (Dushkin 1957: 152). This was proof positive to Ambedkar that his political insights and predictions were correct. In his opinion, the election demonstrated the reality of the threat of upper caste dominance which independence without political safeguards promised the Scheduled Castes. He was also convinced that the Poona Pact was a betrayal of the Untouchables by Gandhi, whose "Epic Fast" was nothing but political blackmail. This interpretation of the Pact and of Gandhi is still widely held by the Jatavs and is a major reason for their present rejection of the Congress Party. To them, the Pact has proven a wolf in sheep's clothing which goes about devouring Untouchables (see page 88). Gandhism has proven not only a tree in whose shade upper caste Hindus can find self-righteous protection, but also a tree whose fruit has grown too bitter for resentful Jatav taste.

Nevertheless, Ambedkar accepted independence as inevitable and decided to continue his fight for the Depressed Classes under the new government, in which the most significant events of his life were to take place. Pandit Nehru chose Dr. Ambedkar to be Law Minister in the first central cabinet of free India, even though the doctor was not a member of the Congress Party. As Law Minister, he was appointed Chairman of the Drafting Committee for the Constitution of India; and events so conspired that he was given the major burden of writing it. Ambedkar guided the Draft Constitution through the Assembly, until on November 26, 1949, it was adopted by the Constituent Assembly. "The poetic justice of having an Untouchable in the role of a modern-day Manu was not lost upon Congress or anyone else, and he received universal praise for the way in which he piloted the Constitution through the Constituent Assembly" (Dushkin 1957: 155). An Untouchable, therefore, was responsible for Article 11 of the Indian Constitution, which declares Untouchability abolished.

Although he sat in the Central Cabinet, Ambedkar's criticism both of the Congress Party and of Hinduism did not abate. He finally resigned from his cabinet post when the Hindu Code Bill, one dear to his heart, failed to receive passage.

Ambedkar's final years were taken up with bitter criticism of the Congress, Nehru, and Hinduism, and with his growing involvement in Buddhism. After his conversion in the fall of 1956, he had little time left to live. On December 6, 1956, he reached his Nirvana. Perhaps the most fitting epitaph on his life was that given by Pandit Nehru, when he called Ambedkar a "symbol of revolt against all oppressive features in Hindu society" (*New York Times* December 6, 1956; quoted in Dushkin 1957: 45). At the funeral pyre in Bombay over half a million people listened to the sounding of "the 'last post,' an honour given for the first time in Bombay to a non-official person" (Keer 1962: 513). The true measure of the man may be found in the effect he has had on his followers and in their enduring devotion to him.

Ambedkar and the Jatavs

Ambedkar's relations with the Agra Jatavs are like a thread of three strands: actual or historical contacts; structural position; and cultural continuities.[2]

Historically, Ambedkar's fight at the Round Table Conferences was known at least to the literate Jatavs. As already noted, a telegram was sent from Agra to London in the name of the Agra Jatavs supporting Ambedkar as their leader and as the representative of their views. They, thereby, rejected Gandhi's claim to be their leader. This support grew to the point that a unit of

[2] Ambedkar's life, his teaching, and his relationship to the Jatavs raise some theoretical problems in the analysis of myth and charisma. These are considered in Lynch (in press).

the Scheduled Castes Federation was formed in Agra in 1942. Support of Ambedkar against Gandhi was a function of three things. First, Agra City has a communications network of railways, post, telegraph, and newspapers. Thus, as urbanites, the Jatavs had access to information about Ambedkar and his movement not readily available in villages. Second, there was in Agra a literate Jatav leadership that could not only avail itself of this information but also evaluate the contents. Third, as mentioned in preceding chapters, the Jatavs had already achieved a degree of independence and freedom of action which permitted them to do much as they liked.

Ambedkar, in 1946 and again in 1956, came to Agra himself and made speeches. He was known by sight to the mass of Agra Jatavs, who felt they had experienced his charismatic view (*darshana*). The arrival of Ambedkar in 1956 was not noticed by the Scheduled Castes alone. An Agra newspaper reports the event thus:

> The awakening of the Jatavs will not be stopped by the purchase of selfish and self-interested leaders. If the Scheduled Castes don't get their full rights, then there will be a revolution. The eyes of all people ought to be opened to the awakening of the Scheduled Castes, especially the Jatavs who were waiting for hours in a great crowd on the Ram Lila grounds for Dr. Ambedkar. It is a fact that such a crowd was not present when Prime Minister Nehru came, since the crowd was sitting packed close together. In such a condition it could not have been less than a *lakh* of people.
> The Jatavs are the natural harbingers or leaders . . . of the Scheduled Castes. Dissatisfaction cannot be eliminated by blinding one's eyes to it or by buying with a little money those selfish and self-interested leaders who for their self-interest would sell their own caste. It can be eliminated when their true leader and the youthful Jatavs become united to fight for their rights and make the government believe that their unity is for fulfilling their duty to the Scheduled Castes (*Sainik* March 20, 1956).

These historical events were strengthened by Ambedkar's structural position. He was an Untouchable, a status he and the Jatavs occupied in common, and which Gandhi did not share. There is a general assumption in Agra that only an Untouchable can really understand and achieve empathy with other Untouchables,[3] and this underlies the often heard phrase that he is "one of our men." In this sense, then, Ambedkar was one with the Jatavs and was structurally identified with them.

Ambedkar had achieved recognition as an Untouchable leader, first from the British at the Round Table Conferences, later as Member for Labour and still later as Minister for Law in the first Nehru Cabinet. He was also an extremely well educated man, holding the coveted foreign degrees of Doctor of Philosophy and Doctor of Science. He was a father of the Constitution of India and a Barrister-at-Law. In these powerful and prestigious statuses he was the most qualified and the highest placed Untouchable in India. He had power that no other Untouchable at the time could surpass. He was, therefore, "our man at the top" for the Jatavs; the man who could best serve their needs and speak for their interests. As early as the 1940s, when he was Labour Member, they asked for and received his help in getting a higher priority rating for railroad shipments of their shoes during World War II. He is believed to be, and indeed was, the architect of the Constitution, which abolishes their Untouchability and grants them full citizenship and the franchise. An informant notes:

Dr. Ambedkar was a great man because he was honest and possessed integrity. He had reached *samadhi*. He was all for his people and only for that. Some say he was popular because he got jobs for us or because he fought for us, but the real secret is that he was all for us with all his being.

Finally, Ambedkar was a revolutionary. He led the fight against Untouchability, Hinduism, and the Brahman caste so hated by the Jatavs. Not only was his interpretation of Gandhism, the Poona Pact, and Hinduism accepted by the Agra Jatavs, but also he gave them a counter-ideology. Thus the Jatavs, who were striving for mobility and who could easily become revolutionaries themselves,

---

[3] I fully realize the irony of this statement and make no special plea for myself. The present work is in itself evidence for the objective understanding of Jatav life and problems that I have been able to achieve.

quickly identified their conflicts and struggles with those of Ambedkar.

In this way Ambedkar's three statuses of Untouchable, national leader, and revolutionary brought him into a structural relation with the Jatavs which made him their choice for paramount leader. In other words, he was a "reference individual"[4] with whom the Jatavs could identify and whom they could imitate. As one informant, a college student, told me, "Parents hold up Dr. Ambedkar as a model to children [and tell them] how he became educated by his own efforts and worked hard. He is worshipped as a god." This model is reinforced by the many songs and poems that have been composed about Ambedkar and are today recited on all festival occasions.

The third strand of Ambedkar's relations to Agra is cultural. This strand is spun of two traditions, the heroic and the religious. The heroic tradition has its roots in the lays and ballads sung in North India called *Alha*. During the rainy season, when work is slow in Agra, many ballads are sung, but the ballad of Alha and Udal (*Alha-Khand*)[5] is always first and foremost. Nowadays, however, along with these ballads the Jatavs also sing ballads and write poems about Dr. Ambedkar and the battles he fought against the upper castes to win freedom from untouchability. Ambedkar, thus, fits into the heroic tradition. He has become another, if not the greatest, culture hero of the Jatavs. He has shown that an Untouchable can be a great leader, a constitution maker, and a warrior in the fight against Untouchability.

The second part of the cultural strand in Ambedkar's relations

[4] "The person who identifies himself with a reference individual will seek to approximate the behavior and values of that individual in his several roles" (Merton 1957: 302).

[5] This is about two brothers, Alha and Udal, who were the warriors of Parmal, ruler of the principality of Mahoba. They waged war and performed many valiant and extraordinary deeds in a long war against Prithiraj, the Chauhan ruler of Delhi in the twelfth century. The origin of the two brothers is uncertain; some claim that they were Rajputs and others claim that they were Ahirs (Waterfield 1923: 9–25 and *passim*). However, the Jatavs of Agra feel that they were Untouchables and, therefore, are Untouchable heroes who have proven the greatness and worth of the lowest castes.

with Agra is the religious tradition. There is in North India a long tradition of saints who have abjured the caste system in one way or another. These saints seem to have been followers of Ramananda, an itinerant follower of the Vaishnavite sect. He taught that perfect faith was perfect love of God and that all servants of God were brothers (Grierson 1919: 569–72). Among Ramananda's disciples were Kabir and Rai Das. Kabir rejected many of the outward signs of Hinduism—its rituals, fasting, and asceticism—but more importantly he rejected caste distinctions and preached that salvation is to be gained by devotion (*bhakti*) (Burn 1919: 632–34). His cult is still celebrated in verse and song. Many Jatavs were followers of Kabir. Rai Das was also a follower of Ramananda and was a Camar. Most of his followers today are Camars. Of the two, however, Kabir is best remembered because of his eloquent hymns and verse. Rai Das is mostly remembered as a caste name adopted by some Camars (see Cohn 1955; Bhatt 1960; Briggs 1920: 210–11).

Ambedkar continued this tradition, because he taught that caste not only was unjust but also was immoral. He established a new dispensation, a new religion (Neo-Buddhism) whose foundation is its unequivocal rejection of Hinduism. Ambedkar set forth his reasons for total rejection in an undelivered talk to the Circle for the Elimination of Casteism (Jat-Pat Todak Mandal) of Lahore in 1936. The speech, later published as *Annihilation of Caste* (Ambedkar 1945), is a mighty salvo against Hinduism which Ambedkar felt was the basis of caste. If caste was to be destroyed, he said, then its religious foundation in the Vedas and Shastras must also be destroyed. Faith in these scriptures was nothing more than a legalized class ethic favoring the Brahmans. "If you wish to bring about a breach in the system, then you have got to apply the dynamite to the Vedas and the Shastras, which deny any part to reason, to Vedas and Shastras, which deny any part to morality. You must destroy the Religion of the Smritis" (Ambedkar 1945: 70).

The superstructure erected on this fundamental doctrine is the

doctrine of the origin of the Buddhists. In 1948 Ambedkar published his book, *The Untouchables: Who Were They and Why
They Became Untouchables* (Ambedkar 1948).[6] The thesis of this
book is that the Untouchables were originally broken men, stray
survivors of the indigenous tribes conquered by invading sedentary
agriculturalists, the Brahmans. These broken men came to live on
the outskirts of the villages as laborers for the conquering agriculturalists. The broken men were hated by the sedentary Brahmans
in the villages because they were Buddhists, and Buddhism was
antithetical to the Brahmanic religion, which at that time practiced cow sacrifice. When the Brahmans saw that they were losing
the masses of the non-Brahmans to Buddhism, which forbade the
slaughtering of the cow but not meat-eating, they were forced to
engage in one-upmanship in order to regain their leadership
from the Buddhists. They gave up all meat-eating, became vegetarians, and declared the cow sacred, which it was not to the
Buddhists. Thereafter, the broken men became Untouchables.
Because of their poverty, they were forced to eat carcasses of the
now sacred cow and by doing so they became polluted outcastes
(see Ambedkar 1948). In this way Ambedkar identified the modern
Untouchables with the ancient Buddhists. He also supplied a rationale for this identification. The Untouchables, descendants of
the Buddhists or broken men are the true autochthenes, carriers
of the true and only valid Indian tradition. Their destiny, then,
demands that they revive their cultural heritage and that they
regain their ancestral patrimony. The Brahmans, on the other
hand, descendants of the invading conquerors of the past, are
the purveyors of oppression, the enforcers of a foreign and alien
tradition. Justice demands that their alien traditions be excoriated from Indian soil and that they return the land to its true
owners.

A well educated and popular leader told me, "Dr. Ambedkar
was a real Bodhisatva. But he was the Martin Luther of Buddhism.
He wrote the bible of Buddhism, *The Buddha and His Dhamma*,"

[6] The book was subsequently translated into Hindi.

a systematic summary of Neo-Buddhism. In this sense Ambedkar has transcended his purely political status of leader and has added to it the haloed status of a saint and a sage (*mahatma*). As one informant told me: "My feeling about Baba Sahab is that he came to us having assumed the body of a messiah, and those things he told us are as true as they are eternal. Baba Sahab was a 'great man' [*mahan*] and by his 'way' all ought to become great."

Because Ambedkar embedded his revolutionary message in religion and in religious myth, its appeal to his followers is far greater than that of his purely political teachings.[7] The justness of this mission lies in its nationalism; it is meant as a messianic commandment for all Indians to save their motherland. This is now referred to in Agra as Baba Sahab's Mission. "At the time of conversion, he told his people that they must be honorable, respectable, responsible Buddhists, and if they could accomplish this, 'We will save our country' " (Zelliot 1966: 205). Ambedkar well knew that religion was suffused into every nook and cranny of Indian society and that the mind of the masses had not yet learned to distinguish that which is Caesar's from that which is God's. He therefore turned to " 'religiofication'—the art of turning practical purposes into holy causes" (Hoffer 1966: 15). Buddhism is a political religion which the Jatav masses can respond to. It pours the new wine of political modernity into the old bottles of religious tradition. It was a stroke of genius, whether conscious or not, for Ambedkar to define the roles of a Buddhist in virtually the same terms as those of a citizen of democratic India. The interchangeability of these two statuses allowed for an appeal to both the traditional and the secular minded among the Jatavs; yet, both could work for the same revolutionary goals.

In short, Buddhism is the language of "saintly" politics; the language so persuasively spoken by Gandhi and others when appealing to the Indian mind.

---

[7] Ambedkar was not the first to do this. Around the turn of the last century, Tilak couched his political objective of freedom for India in terms of British interference with Hindu religious practices.

[The] language of saintly politics is to be found "at the margin" of Indian politics. By this is meant the fact that it is in some quantitative sense relatively unimportant, spoken by a few, and occupying a definitely subsidiary place on the political page. But "margin" may also be allowed here to have something of the importance given to it in economics: there may be few or none actually at the margin, but the location of the point has an effect on all operators as a kind of reference mark. In other words, saintly politics is important as a language of comment rather than of description or practical behavior (Morris-Jones 1964: 59).

I would take this a step further and say that for the Jatavs, Buddhism as an association[8] has bridging functions. The *Buddhist* movement might better be considered as the Buddhist *movement*. It has arisen at the point where traditional Hindu institutions and caste inequality are tangent to modern secular institutions and democratic equality. At present, the Indian socio-political system is a mixed system containing elements of tradition and modernity, caste and class. Buddhism functions to bridge the gap between these two systems and to soften the transition from one to the other. It is religious and it is Indian; it therefore appeals to the Jatavs' need for stability and roots. Yet, it is also secular and egalitarian; it therefore appeals to the Jatavs' need for change and new goals.

It becomes readily apparent why Dr. Ambedkar, a man of the Mahar caste from the state of Maharashtra, not only was accepted but also was apotheosized as a Bodhisatva and made the culture hero of the Jatavs in Uttar Pradesh. His status of Untouchable made him one with the Jatavs. His structural position as a national leader made him their natural choice for leader of the Untouchables. His status as a revolutionary gave him a status similarity with the mobile Jatavs. His anti-caste religious teachings gave him a cultural continuity with a traditional past that is familiar to and valued by the Agra Jatavs.[9] The Jatavs are now counter-

---

[8] My understanding of associations arising at the point where two institutions are tangential to one another is based on Chapple and Coon (1942: 416–23).

[9] The tradition of the equality of man seems to have been especially strong among Camars. Mahar (1958: 62) notes, in her analysis of four leaders of Khalapur village

moralists. They have rejected the Hindu caste ideology of the majority and have put forth a counter-ideology which they believe to be not just a better creed; it is more. It is, in their eyes, the ideology of the future for all India.

Buddhism in Agra

When Ambedkar came to Agra for the second time, in 1956, he told his audience of

his wish that in this year [1956] I will become a Buddhist. Showing the Buddhist religion to be more auspicious than other religions, he said, "I hope you will become Buddhists but this is up to you, therefore I won't press you to do so. However, as soon as I convert to Buddhism, I will not remain as a Scheduled Caste. But you should remember that the reserved seats are only for ten years. They will end soon. . . . In the end you will have to depend upon your own strength" (*Sainik* March 19, 1956).

A counter challenge to this appeal was thrown out by Mr. Jagjivan Ram, the Congress Party leader of the Untouchables and at that time Congress Minister for Railways in the Central Cabinet. In a speech given in Subhash Park, Agra, a month after Ambedkar's appeal, Jagjivan Ram said: "I can't change my religion. He who converts is a coward. Such people can't do any service for the Jatav community" (*Sainik* April 16, 1956). With the alternatives so clearly presented, the Jatavs chose to follow Ambedkar in the path of alienation both from Hinduism and from the Congress Party.

Ambedkar died on December 6th of that year. In death he has become, perhaps, more powerful than in life, for he has become a saint in a hallowed North Indian tradition. On January 13, 1957, a solemn procession was held in Agra. Ambedkar's son, Yashwant Rao, and other leaders came bearing a silver urn containing the doctor's ashes. The urn was installed in the new Buddhist temple (Bodh Vihara) at Chhaki Pat in Agra. This temple, established

---

in North India, that only the Camar could conceive of a unity overriding Untouchability. See also Cohn (1955).

in Agra on March 18, 1956, was a manifestation of the readiness of some Agra Jatavs to follow Ambedkar in his conversion. At the Agra obsequies speeches were given against Hinduism (it was called a worm eating away at society and corrupting it) and an estimated 3,000 were converted. In the 1961 Census 2,262 people (India. Census Commission 1963: 40) were listed as Buddhists in the urban areas of the Agra district and almost all are Jatavs from Agra City.[10] While this is a very small percentage of the total Jatav population, it is actually a very large number when one considers the difficulties of an openly declared convert (*infra*) and the fact that there were virtually no Buddhists in Agra prior to 1951.

After this mass conversion, a revolutionary or conversion campaign (*krantikari daur*) was initiated, and on the next day twenty to twenty-two Hindu temples were converted into Buddhist temples by the Jatavs. In a few cases there were objections from Hindus of other castes, but on the whole it went peacefully.

A second major temple has been established in a place called Diggi. On Sundays, dedicated Buddhists meet there to pray and to read selections from Ambedkar's book *The Buddha and His Dhamma*. Programs are also held there on major Buddhist festivals.

After the conversion an organization called the Indian Buddhist Society (Bharatiya Bodh Mahasabha) was formed but for all practical purposes it remains a nominal association. Its president organizes the Buddhist festival programs and publishes from time to time a newspaper containing articles on Buddhism. His shoe factory in the shoe market, however, is a meeting place for all Buddhists in Agra; it is this location more than any of the temples that forms the center of Buddhist organization and communication in the city.

Mention has already been made of obstacles in the way of the Buddhists and the Buddhist movement. The first, and probably the most significant, of these difficulties is the fact that part of the

---

[10] The Census Report also listed 44 Buddhists for rural areas of Agra District. The impact of the movement is, therefore, centered in urban areas of Agra District.

definition of a member of the Scheduled Castes is that he is a Hindu.[11] Therefore, one who openly professes to be a Buddhist is not eligible for the special help given by the government to the Scheduled Castes under its "protective discrimination" policy. A Buddhist in Uttar Pradesh[12] cannot stand for reserved seats in an election, nor can he apply for government jobs reserved for Scheduled Castes. His children cannot claim eligibility for remission of fees, grants, and scholarships given to the Scheduled Castes for educational purposes.

In such circumstances, then, the Jatav will engage in "bridge actions." Vis-à-vis the state or nation he will activate his Scheduled Caste status, which requires that he be a Hindu, since this carries with it the benefits of "protective discrimination." However, in situations in which he is not facing a government official, he will activate his Buddhist status. This is not to say that the Jatav does not resent and deny the state-required and state-imposed controlling status of Hindu requisite to the claim of Scheduled Caste status. Indeed, one of the primary demands of the Republican Party is that Buddhists be allowed to claim the rights of the Scheduled Castes without being Hindus.

The Jatavs themselves recognize that conversion has not changed their economic and social position in Indian society among other castes. Yet there is some hope that it will change. Nonetheless, if the Buddhists were allowed the benefits of the Hindu Scheduled Castes, there seems little doubt that the number of declared Buddhists in Agra would grow enormously and the practice of Buddhist ritual, and so on would become widespread and public. There is some indication of this: in December 1964 and in January 1965 the Republican Party, again demonstrating its complemen-

[11] "The courts found this religious classification a reasonable one in view of the fact that Scheduled Castes are intended to include those who suffer under the stigma of disabilities of untouchability, a condition which supposedly exists only within Hinduism and which change of religion supposedly effaces, at least in part" (Galanter 1961: 63).

[12] In Maharashtra the Buddhist Mahars have been granted Backward Class status, which allows them some privileges of "protective discrimination," though not as many as those given to Scheduled Castes.

tary relationship with the Buddhist movement, launched an act of civil disobedience. Involved were about 300,000 volunteers, mostly from Maharashtra, Uttar Pradesh, and the Punjab. One of its demands was the retention of Scheduled Caste privileges by Buddhist converts. The act of civil disobedience was brought to a halt only by the assurances of the then Prime Minister, Lal Bahadur Shastri, that the demands would receive favorable consideration (Zelliot 1966: 193).

A threat of Hindu retaliation has placed a second obstacle in the path of Buddhist expansion. In 1956, a large number of Buddhists in the neighboring city of Aligarh organized a protest movement against upper caste Hindus who had objected to the conversion of a Hindu temple into a Buddhist temple (*Nau Jagriti* May 15, 1957). When the protest got out of hand, many Buddhists were imprisoned. The Jatavs of Agra have a little fear, and no little hatred, of similar Hindu retaliation.

The third difficulty in the way of Buddhism's growth in Agra is a lack of organization, leadership, funds, and time. There was great internal dissension over who should be officers in the beginning of the movement, and leadership continues to be ineffective and weak even today. Many Agra Buddhists have expressed the desire that foreign Buddhists should come and instruct them in the faith, but the ship's sail of these missionaries is nowhere visible on the horizon, and it seems unlikely that it soon will be. Attempts to train their own Buddhist monks have proved abortive. Furthermore, effective leadership and energy of the caste are, as we have seen, directed to politics rather than to religion. The Republican Party with its concern for bread and butter issues is of more immediate and practical relevance than is Buddhism with its concern for idealistic and utopian issues. In short, the success of the Party has somewhat quenched the thirst for the religion.

Finally, the Jatavs are, on the whole, more Ambedkarites than they are Buddhists. It is more important that they follow and revere him as their culture hero than that they follow the Buddha and his noble truths. Indeed, there are few Jatav homes without one or more pictures of Baba Sahab, either as Father of the Consti-

Elephant (at top) upon which an effigy of Ambedkar (below) is carried
during the festival of Ambedkar's birthday.

tution or as a Bodhisatva. He is more celebrated in song, legend, and verse than is the Buddha.[13]

There are four festivals in which the praises of Ambedkar and the Buddha are heard. The first is Ambedkar's birthday (*Ambedkar Jayanti*) held in April; the second is the Buddha's birthday (*Bodh Jayanti*), held in May; the third is the Day of the Emperor Ashoka's Victory (*Ashok Vijaydashmi*); and the fourth is Ambedkar Memorial Day, the day of Ambedkar's death. These festivals are celebrated only by Jatavs in Agra, though occasionally a sympathetic member of another caste is invited to give an address at one of the festival programs.

Ambedkar's birthday celebration is held on April 14. It is the largest and most important of the Buddhist festivals in Agra. Among the thousands that attend the fair held on this day are Jatavs from periurban villages and some Mahar soldiers from the nearby Agra army station.

The focal point of the day is a seven-hour parade which begins about eight o'clock in the evening. In it are imaginatively decorated floats, sponsored by the various Jatav neighborhoods in the city. Most of them depict a famous incident in the life of Ambedkar or the Buddha. The parade's main attraction is an elephant which carries a seated, life-size image of the doctor. The parade winds its way through the streets of Agra, announcing to one and all that the Jatavs and their hero are an important presence and ought not to be taken lightly.

On the next day at Nala, an Ambedkar Memorial is held. Fewer people attend this event, but it still attracts a large crowd. Speeches are given by local Jatav leaders and members of the Republican Party. Typical of these is the one paraphrased below. It pulsates with the political awareness of a man who is literate but not well educated, and in this sense is a microcosmic summary of the Jatav political mind.

---

[13] Ironically, Ambedkar himself foresaw and warned against this type of hero-worship, which he felt was a peculiar weakness in India. He said, "*Bhakti* may be a road to the salvation of the soul, but in politics *Bhakti*, or hero-worship, is a sure road to degradation and eventual dictatorship" (quoted in Keer 1962: 412).

The members of the Legislative Assembly want more salary while we want more food. They get money from us and also take the money meant for us that comes from abroad. We will use the method devised by them [the Congress and the Brahmans] against the British, against them themselves. Nehru's government promised us 18 per cent reservations in government jobs and up to now they haven't filled it. Only the sweepers' job is filled [a job which only a sweeper caste will do in India since it involves contact with human faeces]. Let them take up the broom and we their work. Congress promised bank nationalization but they really don't want it because then all their accounts would be shown in public.

This man spoke so evocatively of his audience's sentiments that they would not let him stop for a second breath.

The second festival is Buddha's birthday. This day is now celebrated only by the stalwart Buddhists. It is interesting to note, however, that in 1959 when the 2,500-year celebration of Buddha's birthday was held throughout India, there was a big celebration in Agra. The local newspaper *Sainik* devoted an entire issue to it, and a celebration was held in which speeches were given by all the leading people of Agra. The Jatavs held their own celebration at the same time at the Buddhist temple in Chhaki Pat. Even a second but separate parade was held (*Sainik* September 15, 1958). This separatism indicates the strength of Jatav feeling that they are the only true Buddhists in Agra and that Buddhism is a protest movement against Hindus. The festival was nationally celebrated to underline the fact that India was the birthplace of Buddhism and therefore worthy of world attention and publicity. A year later, on the same day, I found no mention of the occasion in the same paper, *Sainik*. Nowadays few Jatavs attend the ceremony, though all know of it and recognize it as their own.

The third festival is Ashoka's Victory Day, which is held on the same day as the Hindu festival of Dushera, generally near the end of October or early November. The Buddhist temples at Diggi and Chhaki Pat are decorated with flags and fluorescent lights. Many people, especially women in groups, come to Chhaki Pat on this day to worship before the image of the Buddha before going on to the Hindu celebration held on the Ram Lila grounds where a huge effigy of Ravanna (a mythical king of Ceylon) is burned. In the climax of the Victory Day celebration, the Jatavs burn a similar effigy of a Brahman, called the "Ghost of Untouchability." This is a symbol of Jatav hatred for Brahmans; as true Ambedkarites, they feel that the Brahmans wrote the Hindu scriptures which deracinated the Untouchables. Speeches about Buddhist religion (Bodh Dharam) and its message for the Scheduled Castes match point for point the flames that consume this hated victim. All Jatavs know about this immolation and, even if they do not attend, take delight in it, because for them it symbolizes the destruction of Brahmanical tyranny.

The fourth and final holiday for Buddhists in Agra is Ambedkar Day, the day on which Ambedkar's death is commemorated. On this day the remains (*phul*) of Dr. Ambedkar are placed on a decorated table in front of the Buddhist temple at Chhaki Pat. During the day and especially during the evening, many come to place garlands and make obeisance before the hallowed ashes of Ambedkar. Then, at about five o'clock, the leading members of the community, as well as dedicated Buddhists, gather round. In 1963 I estimated about 500 people were present. A program of commemorative speeches lasts until about eight o'clock. These dwell upon the idea that Dr. Ambedkar has given a way and a plan for political, social, and educational advancement. The speeches also ask the Jatavs to take to this way by putting the plan into action. More importantly, the politicians are openly criticized from this religious platform for not faithfully and selflessly performing their duties toward the community. Buddhism is, as I have said, a language of "saintly" politics.

The four festivals described above occur annually. However, on November 17, 18, and 19, 1963, an event of great importance to the Buddhists occurred in Agra; this was the first All-India Buddhist Conference (Bodh Sammelan). Neo-Buddhists from the Punjab in the north to Mysore in the south came to Agra for the Conference. Not only did they come by train but many came in chartered buses, a clear indication of transportation growth in India and its use by lower castes.

Preparations for the Conference had begun four months before, in July. Agra was chosen for the first conference, it was said, because it was a place dear to Ambedkar; because the Buddha was once supposed to have passed through the city; and because the first Buddhist temple was established there on March 18, 1956. The inspiration for the event came in the person of a Buddhist monk who was probably from Nagpur in Maharashtra, where Buddhism enjoys a vigorous existence among the Mahars. The Conference was probably the first attempt to deepen the Buddhist faith in Agra and to extend the contacts of Buddhists from outside Agra to those in it. Over Rs. 5,000 were collected to make the Conference a success; much of this was gathered by selling subscriptions to Jatavs in Agra.

The Conference itself took place on the Agra Ram Lila grounds, where there were both open space and a huge permanent stage. The stage was decorated with fluorescent and colored electric lights, and Buddhist prayer flags; an enthroned bronze statue of the Buddha was placed between garlanded pictures of the Buddha on the right and Dr. Ambedkar on the left. Candles and incense were set to burn before the statue. Along the sides of the field snack stands and book stalls were set up. In the book stalls there were three types of literature:

Literature about Dr. Ambedkar and his book, *The Buddha and His Dhamma.*

Popular pamphlets on Buddhism and various anti-caste saints such as Rai Das, anti-caste literature of all sorts, and copies of various Republican Party and Buddhist newspapers.

Serious literature on Buddhism and Buddhist texts such as the *Dhammapada.*

Literature of the first and second types was most popular, especially since it was cheaper in cost and simpler to understand. Just as the written word was communicated in book stalls, so too was the spoken word communicated in the tea stalls. Since delegates such as the Mahars came from as far away as Mysore, the exchange of views (in Hindi) and personal contacts did much to unite members of different castes into a single anti-caste, anti-Brahmanical movement. Jatav and Mahar met here in the same identity of Buddhist. Many Jatavs also invited Mahars to come to their homes for a meal and to see their living conditions. In turn they were told of life in Maharashtra and Mysore. A paraphrase of an interview with some Mahars from Nagpur clearly brings out the impact of these experiences.

Q. Why did you come to the Buddhist Conference?

A. We came here because this is the first event of its kind since the death of Dr. Ambedkar. We wanted to meet with others and to learn more about the Buddhist religion.

Q. What profit did you derive from the Conference?

A. We got much from the Conference. First we were able to help make it a success in numbers and to show that we Buddhists are a large group here in India. Also, we have met with other people from the rest of India and made contact with them. We Buddhists don't believe in discrimination against anyone; all should be equal, and we met with others as equals in this Conference. We have also learned how to be hospitable. In Nagpur hospitality customs are not so great and good as they are here in Agra. We hope to apply this lesson when we have our own gatherings.

Q. You say you have made contact with others. What do you mean by that?

A. People here in Agra have invited us to their homes. We have talked with them and learned their problems and compared them with our own. We have gotten the addresses of people with whom we can keep in contact and invite to our own functions.

Q. What was the most profitable event of the Conference for you?

A. It was B. P. Maurya's[14] speech. He spoke the truth about ourselves. Most important of all, he has made contact with other countries and let them know of our condition as Buddhists: how we are discriminated against as it was in South Vietnam. This will bring the world and other Buddhists to our support.

[14] At that time a member of Parliament, a Jatav, and member of the Republican Party of India, Maurya had just returned from a three-month tour of the United States as a guest of the Friends of India Society.

The last answer indicated that the third, and most important, channel of communication was the speeches. It also shows that the Buddhists want their case to be known by those outside of India and that they identify with Buddhists the world over. Persecution of Buddhists by the Diem regime in Vietnam was resented and a resolution condemning it was first passed at a meeting of the leaders and was then ratified in the general conference. Buddhist monks from Southeast Asia and Ceylon were invited to the Conference. Those who came included a refugee Buddhist monk from Tibet, a monk from Thailand, and small groups from Ceylon and the Buddhist shrine at Sarnath, India. Even the mayor of Agra made an appearance and the deputy mayor, neither a Buddhist nor a Jatav but a Republican, made a speech. That the mayor, a man of merchant caste, did attend is clear indication first, that as voters the Jatavs command some influence in Agra; second, as Buddhists they make claim to a religion which in India merits at least public respect.

The most important part of the Conference was the meetings of the Buddhist leaders, among whom the most distinguished were also political leaders. In these meetings the internal problems of the movement and its external relationship to the Mahabodhi Society of India[15] were discussed.

The need for more educated Buddhist monks to instruct the faithful was debated at great length, though no definite action was taken at the time. The discussion, however, did underline the desire of the rank and file to learn more about the Buddhist religion. The resolutions made at the meeting and then passed by the general conference included the following:

1. That a Buddhist Educational School be established in India.
2. That the Buddhists give full cooperation to the government in defense of the country against China.

[15] The Mahabodhi Society is a group of Hindus and Buddhists with headquarters at Calcutta in India. It is not composed of Neo-Buddhists and has "connections" both inside and outside India. It appears to have tacit, if not semi-official, approval of the government.

3. That the government be requested that Buddhists be represented in the Indian Parliament by presidential nomination, as are other minorities.
4. That Buddhists be given the educational privileges of the other minorities in India.
5. That the Government of India appoint a Buddhist Religion Advisory Board to look after Buddhist pilgrimage places and institutions.
6. That a second Conference be held in Mysore in a year's time. [It has not yet happened.]

On the final day of the Conference the monks (*bhikkus*) administered initiation (*diksha*) to almost fifty new Buddhist converts on the main stage of the Ram Lila grounds. This was done publicly and with dramatic effect.

In the description of the meetings I have noted a two-pronged Buddhist identification; it is both domestic and foreign. The rationale for this double identification with Buddhists of other countries and with Buddhists of ancient India was described by a prominent Agra Buddhist as follows:

Buddha and Buddhism have had a history in India and there is a certain respect for them. Also, there are many Buddhists outside of India. If we claim to be one of them, then we cannot be Hindus and part of a caste system. There is respect for Buddhists outside of India, therefore, why not for us?

This statement is exactly to the point, when its social significance is considered. The key word is respect. "Having prestige, gaining the esteem or enjoying the deference of others, already means being able to command their actions in some respect" (Nadel 1957: 118–19). In other words a latent function of conversion to Buddhism is to gain a higher status in the social hierarchy of India by forcing the esteem granted Buddhism in India to be granted to the new Buddhists also. Just as the Government of India's esteem for Buddhism can be seen as an implicit political tactic to gain international prestige, so too can the conversion be seen as an implicit political tactic to force the government, and through the government other castes, to give the Neo-Buddhists that favorable treatment and esteem it grants to Buddhism as such.

It can now be seen that the conversion to Buddhism was not a quantum break with the past. In addition to cultural continuities (p. 140 ff.) facilitating the acceptance of Ambedkar and his message, there are continuities between the present Buddhist movement and the past Jatavs-as-Kshatriyas movement. First, the structure of the myths is the same. Both take the form of a "back to the origins" legend and seek validation for it by reference to an age when things were different. As in the Parasuram legend, so too in the Buddhist story, the Jatavs' original status was somewhat higher but has fallen because of the Brahmans. Both legends are anti-Brahman. Both legends make a claim to a status higher than the one presently occupied. In other words, both legends give a "thought of model" of the social structure *as it is* in the present, *as it was* supposed to have been in the past, and *as it ought to be* in the future.

There is, however, one significant difference. The Parasuram legend was defensive; it sought only to legitimize and justify a claim to a higher status within the caste society. The Buddhist legend is, on the contrary, offensive; it legitimizes and justifies the Jatav mission to change and reorganize Indian society on an egalitarian principle. This Buddhist ideological offensive finds its sociological counterpart in the activist Republican Party.

Second, Buddhism is part of the great tradition of India. As a religion it is not a foreign import such as Christianity or Islam, and its form and ideas are not completely alien to an Indian. Furthermore, Ambedkar's version of it is particularly suited to the problems of socially mobile Scheduled Castes. This version has been rationalized as true Buddhism. A well-educated monk, now resident in Ceylon, who translated Ambedkar's *The Buddha and His Dhamma* into Hindi, answered me on the question of the new version of Buddhism in the following way:

Q. What do you think of Dr. Ambedkar's Buddhism? He changed
it some.

A. There is no doubt that he had done something to Buddhism.
. . . Yet there are all kinds of Buddhism. In Ceylon it is different from
Thailand, and that is different from Japan. Buddhism finds itself according to the country in which it is. No doubt, Ambedkar emphasized
certain things more than others, but what he has done is to put Buddhism in a form whereby it will be understood by the people of the
country, India. This needed to be done.

Third, the objectives of both the earlier Jatavs-as-Kshatriyas
movement and the later Jatavs-as-Buddhists movement are the
same. Were the Neo-Buddhists considered true Buddhists and
granted the respect that other Buddhists are granted in India, they
would in effect change their controlling status of Untouchable to
that of Buddhist. This would probably put them on a par with
such religious sects as the Jains and the Christians. From such a
position, entrance into the power and opportunity structures
would not be as blocked as it is now. In other words, if they were
considered true, non-polluting Buddhists, it would be legitimate
for them to occupy the positions of wealth, power, and prestige
generally restricted to upper castes. In reality, however, the Neo-
Buddhists are considered parvenus, and their claim to be Buddhists
is not accorded to them by other castes, who, informally at least,
still consider them as Untouchables and treat them as such. Thus,
conflicts arise.

This situation is not peculiar to the Agra Jatavs. It is based on
the structural position of a consciously mobile but low-ranking
group in an opportunity structure whose doors are, at least informally, closed to it. The same condition exists among many socially mobile groups throughout the world (Lanternari 1963:
passim). Berreman succinctly characterizes the situation:

If . . . there is a general disillusionment with the valued dominant
culture or the chances of being successful in it, a combination of out-
group alienation and membership group loyalty may lead to glorifica-
tion of the membership group and overt rejection of the former valua-
tion group even though many characteristics of the latter group have

been incorporated in the image of the membership group (Berreman 1964: 240–41).

Translated into everyday language this means: If you snub me for putting on good manners, it doesn't bother me; no self-respecting person wants the company of a snob anyway.

Buddhism and Hinduism

The ultimate significance of the Buddhist movement for the Agra Jatavs depends upon its relation to Hinduism. They form a curious combination and exist side by side even among the Jatavs.

While almost all Jatavs claim to be Buddhists and identify with them, there can be little doubt that the majority of them continue to be *de facto* Hindus. Except for the confirmed Buddhists, many men and most, if not all, women continue to celebrate Hindu holidays such as Holi, Dushera, Raksha Bandhan, Krishna Jana-mashtmi, and Diwali. On these days various rituals are performed and Hindu deities are worshipped. In almost every home a small light is lit and worshipped at eventide by the woman of the house. In July there is a fair about four miles outside of Agra at the shrine of the god Mahadev[16] in Balkeswar Colony. Many go for the excitement of the fair; but some also go to worship the god. In April, there is a major pilgrimage to Kela Devi (a goddess) in the state of Rajasthan. Thousands of people of many castes, including Jatavs, flock to this spot, which is famous for its powers of granting fertility.

Most Jatavs in Agra appear to believe in rebirth after death of some sort. Some will deny this in one breath, but recount tales of known "certain" cases in another. I have witnessed magic, observed superstitious taboos, and been told stories of evil spirits (*bhut*) of various types as well as of auspicious spirits such as Mohammedan holy men called *pir*. Furthermore, life cycle ceremonies are all in a form of Hindu low-caste folk ritual.

How, then, can the Jatavs be both Buddhist and Hindu at the

[16] A name for the god Shiva.

same time? The problem can be considered first in cultural and then in sociological terms. In terms of culture, Buddhism for the Jatavs is an ideology or a social philosophy. The social philosophy of Hinduism, which embodies and justifies the theory of the four *varnas (varanashram)* and therefore, in practice, the caste system and its inequalities, is rejected by the Agra Jatavs. Buddhism is an anti-caste social philosophy and preaches a utopian society where all shall be equal and none shall taste the bitter fruit of discrimination. Beyond this, it is true that few Jatavs understand even the basic philosophical subtleties of Buddhism. But this perhaps leads them to more readily adopt and more tenaciously hold it. One must remember that Buddhism is really a belief in a millennial future which must be achieved by hard work, sacrifice, and deprecation of the present (which the Jatavs realistically find easy to do). It is, too, a faith revealed by an exemplary prophet,[17] Ambedkar, and is therefore unquestioningly accepted.

The effectiveness of the doctrine does not come from its meaning but from its certitude. No doctrine however profound and sublime will be effective unless it is presented as the embodiment of the one and only truth. . . . in order to be effective a doctrine must not be understood but has to be believed in (Hoffer 1966: 76).

Agra Buddhism is not a system of rituals and symbols. On the other hand, Hinduism is such a system; it is a religion more of rite and ritual than of explicit beliefs. As long as one's social obligations are carried out properly, one is fairly free to believe as he wants. The only explanation I was able to get for most ritual practices among the Jatavs was "because we've always done it this way; it's the way our ancestors did it." It is easy to see, then, how the Jatavs can maintain Buddhism as a system of belief and Hinduism as a system of rite and ritual, because between them there is little overlap. In this sense the Jatavs are unlike the Mahars of Maharash-

[17] For the distinction between exemplary and ethical prophecy see Weber (1964: 46–59). I have classified Ambedkar as an exemplary prophet because he did not seem to *demand* obedience as an ethical duty (see Weber 1964: 55). See also the quotation from his 1956 speech in Agra, page 145.

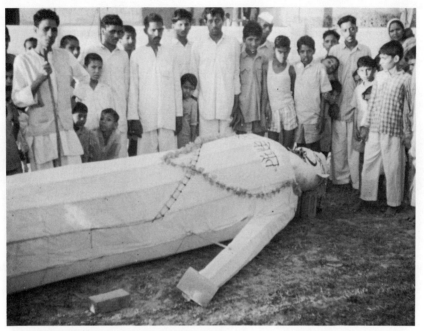

The Ghost of Untouchability, and (at left) Ambedkar's ashes in a covered urn at the Buddhist Temple in Chhaki Pat.

tra, who have attempted to adopt Buddhist rites and rituals as well as beliefs (see Zelliot, 1966). As long as the structural blocks to the spread of Buddhism in the state of Uttar Pradesh remain, this condition is likely to persist.

This duality is not as strange as it may seem at first glance. In fact, it is very Indian. David Mandelbaum has argued cogently that religions of the sub-continent (and many religions for that matter) are characterized by a dichotomy into transcendental and pragmatic aspects. The transcendental aspect gives ultimate meaning to life and its problems, as does Buddhism for the Jatavs; the pragmatic aspect meets immediate needs of its practitioners through magic, rite, and ritual, as does Hinduism for the Jatavs. I would go a step further and argue that the proximity of modern medical services, secular schools, and the pot of gold at the end of the political rainbow have also weakened the hold of pragmatic Hinduism upon the Jatavs and thereby also paved the way for the Buddhist conversion. In Mandelbaum's terms, Buddhism is an "early modern religion" involving, as it does, a rejection of world-rejection (Mandelbaum 1966: 1174–91). It occurs, as I have said (page 144), where tradition and modernity are tangential.

In sociological terms the Buddhists can also be Hindus because they engage in "bridge actions" and status segregation.[18] A man who attends a Buddhist rally at the Buddhist temple finds little difficulty in worshipping Kela Devi in Rajasthan. His religious statuses are segregated in time and place. Yet being Buddhist is especially important to the Jatavs because it forms their reference point of identification and therefore of their place, or rather the place they feel is due them and for which they must work, in Indian society.

It must be reiterated, however, that most Agra Jatavs do not consider themselves Hindus. One often hears in conversation references to "the Hindus," by which is meant caste Hindus who consider Jatavs as Untouchables. It is these upper caste Hindus

[18] Similar concepts are those of Goffman (1959), that is, "audience segregation"; and Singer (1968), that is, "compartmentalization."

who form the Jatavs' negative reference group. "The Hindus" occupy a position considered as "they" and not as "we," and the dividing line of caste status is known to both groups. An informant notes:

We will rebel even though we are the most hated. The Brahmans hate us and we have a thorough hatred against Brahmans. The Brahmans and the Camars are equal in number in Uttar Pradesh, and only fear led to their abhorrence of us. We read the Shastras now only for criticism, not for edification.

Thus we come the full round in this topic and conclude with the assertion that Buddhism in Agra is Dr. Ambedkar. What this means is that it is Dr. Ambedkar who is revered as the culture hero of the Jatavs; it is his name that is on everyone's lips in greeting: *"Jai Bhim"* (Hail Bhimrao Ambedkar), and not *"Jai Bodh"* (Hail the Buddha). In meetings and festivals it is more the deeds of Ambedkar and his teachings that are discussed than those of the Buddha. Ambedkar is the "reference individual" with whom Jatavs can identify and imitate. He has shown that an Untouchable can be great in India. He is the savior who has liberated them by writing the Constitution and who has given them a political party to fight for those rights. He has also given them a rationale with which they can justify and assert their claims for upward mobility. In short, Buddhism in Agra, at least to the present, is the cult of Ambedkar and Ambedkarism.

Summary

In summary, Ambedkar was accepted easily because of his historical contacts with Agra; his cultural continuity with past anti-caste reformers and saints; and his structural position in the nation, which gave him power and structural observability so that he could operate as a true leader and representative of the Jatavs and other Scheduled Castes. Buddhism, too, was readily accepted since it was part of the great tradition of India and not of foreign origin. It has political importance to the Indian government, which seeks the favor of Buddhist countries in Asia, and, therefore, the Jatavs hope to use it as a political wedge to force recognition of themselves. In form and underlying structure Buddhism is similar to previous myths justifying present, past, and future social status. Buddhism in Agra co-exists with Hinduism through segregation of statuses and of ritual from belief. Moreover, it has political functions as "saintly" politics. It is a language of criticism and reform which forms the ideological complement of the Republican Party. Finally, it has integrating functions, because it has united Jatav, Mahar, and others in a single identity overriding narrow and primordial caste loyalties.

Agra Buddhism, therefore, does seem to be another example of what Wallace has called "revitalization movements." [19] He defines such a movement as "deliberate, organized attempts by some members of a society to construct a more satisfying culture by rapid acceptance of a pattern of multiple innovations" (Wallace 1961: 143–44). Such a movement results when: "Under conditions of disorganization, the system, from the standpoint of at least some of its members is unable to make possible the reliable satisfaction of certain values which are held to be essential to continued well-being and self-respect" (Wallace 1961: 144). It has been pointed out that the Jatavs are counter-moralists; they reject Hindu caste ideology because of its legitimation of inequality and of Untouchability. If they were to accept such values, they would also have to accept their own low social status, something they emphatically don't do. Jatav rejection of Hinduism is further enhanced by the ideals embodied in the Constitution and the Five Year Plans. In accepting these they come into conflict with the religious ideals of Hinduism and its social counterpart, the caste system.

Jatav rejection of an old cultural configuration and the concomitant acceptance of a new one is not inconsistent with the total culture that preceded. I have attempted to demonstrate this by showing how the Buddhist break with the past is also built upon many items from traditional culture and society.

---

[19] For a fuller treatment of the Buddhist movement as a "revitalization movement," see Miller (n.d.).

In short, "revitalization movements" may well represent the "rapid acceptance of a pattern of multiple innovations" or a sort of "quantum leap" into a radically new cultural configuration. However, the lines or patterns of such movements seem necessarily to be consistent with the patterns that preceded them (Miller n.d.: 14).

In conclusion a statement must be made about the future of the movement in Agra. There are three possibilities: success, reabsorption, or secularism. First, Buddhism may be successful. In this possibility the policy of the government must be considered. If it allows the Buddhists to receive the benefits of the Scheduled Castes, there is good reason to expect a real change in religious ritual and practice among the Jatavs, leading to a situation comparable to what exists among the Mahars in Maharashtra today. Second, there is the possibility of reabsorption into Hinduism, and perhaps this is the most important. Here exists the problem of whether or not, over a long stretch of time, the pervasive influence of Hinduism will absorb the Neo-Buddhist community just as it did the ancient one. This depends to some extent upon Ambedkar's position in Indian culture and society. Is he just another culture hero in the line of Kabir and Rai Das or is he a new kind of culture hero, whose teachings and message, both political and religious, will enable his followers to overcome the barrier of Untouchability? That will be the true measure of the movement and the greatness of its prophet. And third, there exists the remote possibility that the Buddhist movement might be successful in its bridging function of making the transition from Hinduism to secularism, at least for the Jatavs. The problems here are so many as to make this possibility quite a remote one.

CHAPTER VI

The Neighborhood of Bhim Nagar

Life for a member of an Indian caste is a coin having two faces. On one side a Jatav is seen as belonging to a caste which falls near the bottom rung of the caste ladder, and the Jatav is expected to behave in a certain way when face to face with members of other castes. This is the caste to caste face of the coin and to this point in the book we have been considering how this aspect of caste has been changing in the fields of politics, economics, and religion. On the other side of the coin, however, a Jatav is seen as belonging to an organized group of people who share a common identity, common traditions, and a common meaningful way of life. This is the face of the coin which expresses internal or caste brother to caste brother relations. Our question from this side now becomes: How have the changes in the fields of economics, politics, and religion seeped down and affected the very muscle and sinew of internal caste life itself? To answer this question, we will have to move directly into a typical Jatav neighborhood (*mahalla, basti*) in Agra City; this neighborhood I shall call Bhim Nagar.[1] In such a place, and there are well over two hundred of them generally distributed along caste lines in Agra (Tiwari 1958: 106), one can get some un-

[1] This fictitious name is taken from Bhimrao, the first name of Dr. Ambedkar.

derstanding of what life within the protective walls of caste means to a Jatav. As we shall see, a neighborhood is not merely a place where one lives; it is a distinctive unit of social life as well.

Background

Bhim Nagar is situated between the new, British founded Cantonment, and the old, Muslim founded city. It is bordered on one side by railroad tracks leading to Agra Fort Station and on other sides by houses of other neighborhoods. Lying in a deep gully, it is often flooded and unhealthy during the rainy season, and like most Jatav neighborhoods it is segregated from those of upper castes. Leading off one of the main Agra roads is a small asphalt path which ends in an open field *(maidan)* which dominates the center of the neighborhood. As in many cities outside the Western world, Bhim Nagar is the home of both man and beast; within it there is a dairy with about twenty head of water buffalo and cows which supply milk to it and surrounding neighborhoods. Occasionally one finds a goat tied to a stake or a chicken tethered outside a home. There are also a few packs of dogs whose watchful growls proclaim them guardians more of the neighborhood than of individual homes.

The first settlers to come to Bhim Nagar about a hundred years ago were Shri Ram and his brother Bahadur. They are alleged to have come there in exchange for their land at the present site of Agra Fort Railway Station. Before this time it is said that some potters *(Kumhars)* and Muslims lived on the land of Bhim Nagar. The settlement maps of 1874–75 and 1923–24 show the land as vacant except for a slaughterhouse, which has since moved away. The descendants of Ram, his son, Murli Ram, and his grandson Bhim, were engaged in tanning leather, making bone meal, and scavenging or laboring in the city. But today nearly all inhabitants of Bhim Nagar are shoe makers.

The general appearance of Bhim Nagar does not give firm evidence of the poverty of its inhabitants when viewed within an Indian context. There are 151 brick and cement *(pakka)* homes and ninety-six mud and wood (*kaccha*) homes in Bhim Nagar. The number of rooms for each type is given in Table V. The land on

TABLE V. *Houses in Bhim Nagar*

House Type	Number of Rooms							Total
	1	2	3	4	5	6	7	
Kaccha	65	19	4	3	5	0	0	96
Pakka	55	52	19	11	5	5	4	151
Total	120	71	23	14	10	5	4	247

which many of the homes are built is classified as government land (*nazul*), but its exact status has not yet been decided by the courts. It is for this reason that the earlier settlement maps showed the land as vacant. For all practical purposes the government land is, as the Jatavs feel, theirs by right of heredity or occupancy. Electricity came to the neighborhood in the 1950s and the lamp posts at strategic places are becoming places around which people meet to gossip in the evening. Thirty-four homes have their own electric connections. Only five have private water taps; all others must draw water from the six public taps in the neighborhood. There is also a well built by the people themselves, but it is no longer used for drinking: its water is bitter and the sweeter tap water is preferred. Water taps frequently are the scene of arguments when tempers boil over whose turn is next.

Bhim Nagar now boasts fourteen radios, and a few of these are in public enough places to be heard by those interested in listening. This is important since it not only brings Jatavs into immediate communication with all important national news, but also into the almost nationalized cult of the popular movie song.

Most of Bhim Nagar's footpaths are now paved with brick and alongside of them are little gullies for drainage and for relief of the more basic needs of little children. These are periodically cleaned by the municipal sweeper and are, on the whole, kept well. My own impression was that Bhim Nagar received more sunlight and had more open spaces than many upper caste neighborhoods in older parts of the city. For this reason it was better ventilated and seemed a healthier place in which to live.

Few homes have their own latrines and most people must use the public latrines provided by the municipality on an open space at one side of the neighborhood. It has separate concrete-walled sides for both men and women, but no roof for protection from rain and sun. Frequently, a queue lines up for the available places in the morning; it is an experience few enjoy. Some men and women prefer to get up before dawn and use the side of the rail-road tracks instead. On the outside wall of the latrine is a garbage dump which is periodically cleaned by the municipality. Both the dump and the latrine attract the pigs of the sweepers and clouds of flies. Conditions such as these make many Agra Jatavs feel that life in a village is more pleasant and healthy.

The total population of Bhim Nagar as a neighborhood of the city is 2,058 individuals of whom 1,128 are male and 930 are fe-male; in other words, approximately 122 males for every 100 fe-males. This population is, however, not homogeneous in terms of caste. There are 188 residents of other castes as well as 215 Jatavs who are socially affiliated with other neighborhoods. Thus, the total population of the social unit Bhim Nagar is 1,655 residents of whom 904 are male and 751 are female; a male to female ratio of 1.2: (A more detailed breakdown of the population is given in Table A–III of the Appendix.) Throughout the rest of this book, when we refer to Bhim Nagar we shall mean the social unit, not the place called Bhim Nagar.

Interaction between the Jatavs and members of other castes in Bhim Nagar seems to be confined to specific circumstances. These other castes tend to live on the edge of the main Jatav settlement or in *bakals* (houses grouped around a central courtyard with one main entrance) of their own. The Sindhis (immigrants from Sind) own the largest *pakka* (brick and cement) house in the neighbor-hood and came there after Partition. One of them owns the dairy located in the neighborhood and he sells milk. (A sad commentary on the poverty of the Jatavs is that out of the estimated production

Bhim Nagar. At left, the Buddhist Temple. At the right, a *bakal* (see page 169). Women separating leather scraps (below).

The dairy in Bhim Nagar (above), and (below) washing pots at one of Bhim Nagar's six water taps.

of 80 quarts a day only 10 quarts or one-eighth of the total production is consumed within Bhim Nagar.) Another Sindhi operates
a small provision store and is a money lender. Interaction with
them tends to be confined to times when these items are needed.
The Kolis keep to themselves, although one of them has a small
tailor's shop just outside of the entrance to his *bakal*. The single
merchant family operates a small sweets store. The Muslims live in
one corner of the neighborhood and tend to keep to themselves,
although their small children may play together with Jatavs. The
washermen and the sweepers live on the outskirts of the neighborhood, and they too keep to themselves and their businesses.

Family Life

The earliest remembered facts about family life in Bhim Nagar
go back to about 1920. At that time Bhim Nagar was composed of
100 Jatav families or households (*ghar*). Today there are 270. To
a Jatav a household (*ghar*) is not just a house; it is more appropriately the nuclear or joint family[2] that lives in a house (*dehari*, literally door lintel). Were a joint family to split into two separate
families, there would be two households living in one house. It is
not unusual to find in the courtyard of a house two or three ovens
(*culha*)[3] belonging to brothers or even cousins who have fissioned
out of a single joint family but who continue to occupy their
portions of the ancestral household.[4]

There were in Bhim Nagar at the time of our census in the

[2] A joint family is an extended family. The term joint, however, is generally used
in the literature on India.

[3] A separate oven, however, does not necessarily indicate a separate household, since
allowing wives to cook separately in a joint family may keep the family together for
some time more.

[4] There are two ways of identifying households aside from an actual count. The
first is to observe the sweepers who come to clean the streets, garbage in front of
the house, and latrines. The sweeper will always demand payment according to the
number of households in a house. The sweeper is paid flat bread (*roti*) per oven or
household within a house. The second way is to note in the marriage payment book
(*nyote ki kitab*) who are the individuals and/or families subsumed in a payment
due at marriage. There is a chance of error in this method, however, since two
newly separated households may briefly continue to pay this obligation jointly.

spring of 1963 193 nuclear families; 66 joint families; and 10 single member families.[5] Of the joint families those of the lineal type were a majority with 38; while the collateral type was a minority with 15.

Joint families among the Jatvas do not appear to be of the stable and persisting type. My informants almost unanimously said that after the death of the father, married brothers and their families will inevitably split, especially if the mother has also died. The reasons given for the breakup of joint families are: quarrels between wives; quarrels over use of money; preference to live separately. The basic factor behind these reasons is economic. Where there is a dispute over the allocation of monetary resources, arguments that lead to a split will arise. These again relate to the fact that there is no pattern of primogeniturial inheritance in this caste or in much of India. Thus, while the joint family remains an ideal pattern for family life, its real existence among the Jatavs is contingent upon the non-activation of the rule of inheritance which allocates equal shares in the family estate to all brothers.

In splitting a household, after the death of the father, all brothers have a right to an equal share of the house and its belongings. This is by a mutual arrangement. A daughter gets nothing, but it is the responsibility of brothers to see her married, if she is not married already. The eldest brother succeeds to any title the father may have had and also takes care of any younger brothers and sisters. In the case of one large and prosperous family which had split, one brother set up his own house at some distance but maintained a locked room in his father's house to validate his share in it.

In Jatav families the relationship between husband and wife is one in which the wife is expected to show respect and deference to her husband. A man is allowed to strike his wife but she cannot strike back. A husband can scold his wife but she ought not talk back, though she may privately do so in an indirect way. It is often

[5] For a fuller breakdown and for definitions of family types see Table A–IV in Appendix.

said that a wife must treat her husband like a god. In reality a wife's job is to keep house, cook meals, fetch water, clean kitchen utensils, and bear and raise children. At times she may help her husband by sorting tacks, nails, and leather scraps and by making heels for shoes. The act which best expresses and even ritualizes the relation of husband to wife is that she will not eat her evening meal until he has eaten, no matter how late he may return. This act daily symbolizes the subordinate role of wife to husband, and it was the act most mentioned to me by my informants as manifesting the relation of wife to husband.

Married women are expected to practice *ghunghat* (lowering the veil of the *sari* to cover the face). A newly married girl is expected to do this completely so that the whole face is covered before any man of her husband's neighborhood, since he is like a potential husband to her. As time goes on, however, with age and children she is expected only to keep the veil over her head, and is allowed greater freedom in speaking to other men of the neighborhood into which she has married.

The relationship of son and daughter to their father has been characterized by respect and obedience. Fathers traditionally gave orders to sons, and it was never otherwise except in old age, when fathers could abdicate their authority position but not their right to respect. This same relationship also has been held toward the father's brothers. An eldest son is and was often treated with indulgence by his father, who prepares the boy to take his place. Furthermore, this first son continues to give his father the prestigious status of father in a cultural sense within the community. A man without a son or sons is considered unfortunate.

Education is changing to some extent the roles of father and son. The educated son is often better able to handle business and legal affairs than is his father, and the father will often defer to him in such matters. (A group of men in Bhim Nagar once commented on a father not allowing his adult married son to go somewhere and said that the father "thinks in the old ways.") It is also felt that an educated woman ignores tradition more than an un-

educated one, and this concurred with my observations. An educated woman is more likely to be careless about or not practice
face veiling and would more freely talk with a man. A woman
speaker of Mahar caste from Maharashtra was at the first annual
Buddhist Conference in Agra. She urged Jatav women to drop the
custom of face veiling and to get more education. One "emanci-
pated" young lady who had been educated until the eighth grade
refused to observe this custom; consequently, her husband's family
would not accept her. However, she finally had to accede to their
wishes as well as to pressure from her own family. If the Buddhist
movement ever receives full public recognition and respect, it
would give both legitimacy and impetus for these changes in custom to take place.

The relationship of son and daughter to their mother seems to
have changed little. This relationship is one of subordination and
respect, although its psychological correlates are love and devotion. A mother's control over the adult son is usually indirect but
firm. Two incidents will reveal the control a mother can have over
her son.

A young man of 24 has a successful shoe business. He must occasionally travel to Delhi and other places. He told me he could never go
for a trip longer than two weeks because his mother said she would die
if she did not see his face.

Another young man won a contest for making shoes. As a prize he
would be sent to Russia for a few years to take advanced training. His
mother would not let him go for such a long time.

Marriage is the central event in the life of the individual of the
Jatav community. It is also more than this, since it is the event of
most importance to the internal social life of the caste. It is the
event which is most ritually elaborate and, since Jatav society is
endogamous, it is the rite and event which most symbolizes and
reinforces the internal unity of the caste on the one hand and its
external separateness from other castes on the other.

Marriage is necessary for a girl and her father is obliged to see
that her marriage is properly performed. In one case, a girl was of

weak mind. The father of the girl found a poor village boy who was being married to a girl in Agra City. The father arranged to have his daughter go through the critical part of the marriage ceremony, called circling the bhamver (*bhamver phere*), before it was done with the other girl who was really being married to the boy. The boy's father was given a dowry for doing this and the boy, according to the marriage contract, has no further responsibility for the girl of weak mind. People say that something terrible would happen to a girl after death were she not married.

All marriages (*shadi*) among Jatavs are arranged by the parents or other male relatives of the boy and girl and, as already mentioned, virtually all marriages are within the caste; that is, they are endogamous. Traditionally, the boy and the girl had little or no say in this arrangement and most still prefer it this way. They say, "Our mothers and fathers love us and have much experience; they therefore will choose the best mate for us." The major financial burden of a marriage falls upon the girl's side, because a dowry must be given and the marriage must be performed in her village or neighborhood. Generally the boy's family will stay at the girl's home for two or three days and during this time they must be wined, dined, and entertained. Because of these expenses and because after the marriage the girl's family must treat with respect and deference her husband's family, a girl child is sometimes considered a burden. There is a saying about this: "On the birth of a girl one must bow his head." (*Larki hone se sir jhhukna parta hai*). It appears that child marriages were much more frequent in the past than they are at present. Even in such marriages, however, the consummation did not and does not take place until adolescence when a separate ceremony called consummation (*gauna*) is performed.[6]

Among many college educated men there has been a definite change in attitudes about marriage and marriage partners. Four

---

[6] Much confusion and misunderstanding of Indian child marriage has probably been created by confusing the marriage proper (*shadi*) with the consummation of the marriage (*gauna*).

Jatav college graduates felt strongly that their career had been severely hampered or ruined because of their early marriages. They said it was difficult to study once one had a wife and child. Either parents do not understand their desire for learning and put pressure upon them to contribute to the support of the family, or it is necessary for them to work to support their own family. It is also difficult to get an illiterate wife to conform to new standards of cleanliness and so forth. Thus, some feel that wives are more of a burden than the helper they no doubt still are for uneducated shoe makers. Most educated men prefer a girl with at least primary school education. In the words of one student:

I was married at the age of twelve and three years later I came together with my wife. Now I realize this was a bad thing, but then I was very happy that I was going to be married. Now I have a son. It is difficult for me to be married and to be a student. My wife, though she is from a village, realizes this, and she tries to help me as much as she can.

This student now works during the day and goes to law school at night. He considers himself lucky that his wife supports his ambitions rather than complains about his not earning enough. Another student says: "I am not married yet nor is my older brother. Uneducated people marry very early. Educated people marry late." This student shows that new standards concerning the proper age of marriage are entering the caste. At one level these standards are related to Western ideas and the law that forbids child marriage. At another level, however, they are related to economic expectations. It is easier to complete higher education when unmarried and therefore to get a better job. Thereafter it is also easier to get a better start in life and avoid children that are not yet wanted.

A much more general change is the raising of the permitted age for marriage. There is pressure in Bhim Nagar as well as in other neighborhoods in the city to conform to the law that girls may not marry until sixteen years of age and boys until they are eighteen years of age. Some parents have also come to realize that a boy with a B.A. degree can command a higher dowry than an uneducated boy. I was once told that there was a Jatav boy in Agra with an engineering degree who demanded a girl who could bring a dowry of Rs. 6,000. He was supposed to have claimed that this would repay the cost of his education. Whether or not the story is true, it indicates the perceived correlation between amount of education and amount of dowry. Educated young men are also beginning to exercise greater control over the choice of their mate than was formerly allowed. They are often consulted by their parents or asked to pass judgment on the girl. This is especially true when parents have been influenced by Buddhism, the Arya Samaj, or Western ideas.

Buddhism has also influenced the general attitude toward age of marriage. One uneducated but confirmed Buddhist put it this way: "The Brahmans gave us such a philosophy that we married early and had many children early. Therefore, we became helpless to make progress since we had to provide for a family." I asked this same man what people would say if his own son should be twenty years old and not married. He replied, "They would say your son is so big and not married [in tones of derision]." Nonetheless he said he would marry his son after the boy had completed high school. This man also opposed the dowry system (as did most Buddhists), and was even willing to allow intra-neighborhood marriage.[7] He said, "Marriage should be this way: if you have a sister, and I a brother, and we know each other's character and our siblings' character, then we should make this marriage." The significance of the statement is not that he would actually do so, but rather that he perceives an alternative system he can positively value. There have been some Buddhist marriages in Agra. These are done with simplicity, economy, and speed. However, none took place during the time of my field work. I attended one marriage in the city in which the father of a boy who had a B.A. refused to take a dowry for his son. He believed that the system should be eliminated. I suspect he was a Buddhist.

Education and improved economic status have also influenced the distance at which marriages are contracted in two ways. In

---

[7] Neighborhoods are exogamous. *Vide infra* page 189.

the first place, it is somewhat more difficult to find a mate of comparable education or wealth within Agra, and one must look further afield. Secondly, the educated generally are in government service and are posted to different areas throughout Uttar Pradesh. They are often asked to look for a match either in their home place or at their post. Thirty or more years ago Agra Jatavs who had migrated to Delhi would not marry with the other Jatavs in Delhi. This is no longer true, and there are now a number of marriages between Agra and Delhi Jatavs. Quite definitely there has been an increase in what M. N. Srinivas would call the horizontal solidarity of caste.

In taking a census of Bhim Nagar, I asked the heads of families to tell me in what place their daughters had been married. Thus, in a total of 120 such marriages about which I have information, the following results were obtained:

Agra City	71
Agra District	28
Farukhabad District	2
Aligarh District	1
Etawah District	2
Mathura District	8
Delhi	3
Bharatpur District	3
Bulandshahar District	1
Jamshedpur City, Bihar State	1
Total	120

The results clearly show a preference first for marriage in Agra City, next for marriage within Agra District itself, and thereafter for districts close by, especially for those in the north of Agra such as Mathura, Aligarh, Bulandshahar, and Delhi. Bharatpur is in Rajasthan State, and Agra Jatavs have some ancestral contacts there. The radius of this marriage circle is about 120 miles with the exception of the single marriage in Jamshedpur in faraway Bihar.

Under the influence of Buddhism and the Republican Party the possibility of inter-caste marriages now exists. One wealthy man of Agra has already married his son to a girl of Kuril caste (also Camar) from Kanpur. A politically important Jatav from the neighboring city of Aligarh is trying to marry his niece to the nephew of another politician of the Mahar caste from Maharashtra. The Aligarh politician's relations object to the marriage, because they feel the Mahars are lower than they; he, however, feels his wish will prevail. In view of the Buddhist identity shared by Jatavs and Mahars, such a marriage, if imitated, could have manifold consequences.

In a respondent group of ten men I asked: "Do you support intercaste marriages? [8] Why?" Eight men were in favor and two were against. Those in favor gave such reasons as, "Because then the walls of differences will fall down"; "Because then we will be equal with all"; "Because then we'll sit with them and it will help to raise us up. All will be equal"; "In the past it was like this. It will destroy Untouchability." Of the two against, one said, "I don't like it." The other, an Arya Samajist, said, "There are differences in custom and people treat girls from a lower caste badly. . . . Quarrels would arise." However, he would admit its possibility among those who are called Camars, such as Kurils, Reghars, and Jatavs. The Jatavs' attitudes toward marriage with different castes are tabulated here.

	Sweeper (Bhangi)	Merchant (Baniya)	Priest (Brahman)	Muslim	Tailor (Koli)	Leather Worker (Kuril)
Yes	5	7	7	3	6	7
No	5	3	3	7	4	3

	Christian	Barber (Nai)	Leather Worker (Reghar)	Warrior (Thakur)	Washer-man (Dhobi)	Iron-smith (Lohar)
Yes	3	5	6	7	5	5
No	7	5	4	3	5	5

[8] Jatavs feel strongly about intercaste marriages as such, be they by boy or girl. The question was directed to this rather than to attitudes about hypergamy or hypogamy.

What is evident from this tabulation is the greater acceptance of marrying up, such as with merchants, Brahmans, and Thakurs, than there is of marrying with castes which are sometimes considered lower, such as sweepers, washermen, barbers, and ironsmiths. There is also a feeling against marriage with Christians and Mohammedans, although there are known cases of Jatav-Christian marriages living in Agra today. This feeling is possibly due to the fact that the children of such marriages may reject Jatav status and become Christians. There is also a feeling that marriage of Kurils and Reghars with Jatavs is not only more possible but also more probable than other such marriages, because they too are Camars. There is a definite feeling against marriage with sweepers and the "yes" answers were qualified with statements such as, "if the boy is able," or "if he is educated and has other qualifications." Such marriages are, in Jatav eyes, marrying down.

A final, still incipient change is correlated with education; it is the desire to simplify marriage rituals. Two young school boys in my knowledge refused to go through the ceremony of rubbing on tumeric (*haldi*) before marriage. Another refused to have a brass band at his marriage, wanting it to be as simple as possible despite the fact that he was a postgraduate engineer. Another marriage, which a high school graduate arranged for his younger brother, was carried out in such a way that all the ceremonies took place in one night. The wedding feast was served at tables with table cloths and crockery rather than with leaf plates on the ground. Present at this feast were many of the elder brother's co-workers from other castes such as merchants and Brahmans. Members of other castes came to wedding feasts in the families of some of the Jatav politicians. I attended the wedding feast of a young Jatav holding a B.A. Some of his fellow students attended, and among them were two Brahmans, one Kshatriya (*Thakur*) and one merchant. They said, "We are all one"; they rejected caste status in governing their actions. Such cases, while not very frequent, are nevertheless indicators of attempts to live according to new values by members of many Indian castes.

Thoks

Some neighborhoods are divided into sub-units called *thoks*. In Bhim Nagar there are ten such units, a rather high number that is due to an unusual amount of factionalism in the neighborhood. Structurally, a *thok* (literally, a group, an amount or bundle of the same kind of merchandise) is composed of several households under the leadership of one or more head men (*chaudhuries*) whose positions are inherited in one of the member households. Functionally, a *thok* is a unit of social control, commensality, cooperation, and affective relationship.

One becomes a member of a *thok* either by ascription through birth in one of its households or by achievement, through acceptance of its members. An outsider who settles in a place like Bhim Nagar may gain membership in a *thok* on payment of a small fee and approval of the leading men of the *thok*. He must also agree to abide by the rules and customs of the *thok* and its council (*panchayat*). As a member of a *thok* a man may give and receive *nyota* (loans of money in the form of an invitation to a feast) from other members. He may also sit in the *panchayat* or ask it to settle cases for him. *Thoks,* moreover, are not strictly territorial units or places within a neighborhood, because some of its members may live in the places of other *thoks* or even in other neighborhoods. Nor is a *thok* a unit of real kinsmen, although it tends to be so because it is dominated by one or more descent lines.

Each *thok* has its own *panchayat* formed from the *panches* (leading male members or the *thok*). Formerly the *panchayat* would decide all types of questions, and its decisions were binding. Chief among the types of cases it considered were and to some extent still are cases of marriage and illegitimate sex relations; quarrels between members of the *thok;* and making of new rules of behavior for members of the *thok*.

The leader (or leaders) of a *thok* is its head man, a position passed from father to son. A good head man is one who is a good orator, is advanced in age, and is a good leader; he is one who is astute at finding the hidden issues in a case and the acceptable solutions for them. In the past a head man with a forceful personality and a reputation for fairness could often settle a case without convening a *panchayat*. Such men were both admired and feared. A head man's major duties are to keep peace and order within the *thok* and to reprimand offenders. He also officiates at weddings and at other life cycle ceremonies. In the past, more so than in the present, he represented the *thok* and its members in external matters such as in the courts and with the police.

In addition to the head man, there is his assistant, the mace bearer (*chari bardar*), a position also passed from father to son. The mace bearer summons all the men of the *thok* to a *panchayat,* and he asks for order when the meeting gets too loud or out of hand. He is allowed to use force or trickery, such as hiding a man's tools so he cannot work, when members of the *panchayat* are tardy or recalcitrant.

During the past fifty years the position of the head man has lost much of its prestige and power to coerce. This gradual erosion of authority seems to have begun sometime after 1900 when the "big men" began to assume a place of importance within the caste. By the 1920s, when the Jatav Men's Association was in its heyday, the "big men" had become a caste leadership group parallel to the hereditary head men and gradually assumed the functions of problem solving and policy formulating in matters external to the caste. They could do this because they were better adapted to the changing socio-political environment. They also acted as advisors to the caste head men and *panches* in the caste *panchayats*. As one informant put it when describing one of the most famous of the "big men," "He was like a lawyer at the side of the head *panch*."

After independence there was a further dichotomization of caste leadership. A new group, the "politicians," became differentiated from the "big men," who are now largely confined to leader-

Children passing the time of day (above) in one of the *thoks,* and playing holy man (below) after the rains.

Young man does his own laundry.

ship in the shoe business and other financial matters. The "politicians" have taken over in most other areas of leadership. The introduction of parliamentary democracy and the universal franchise were the bases upon which their new status was established. Because they have assumed part of the role-set of the old head men, particularly in caste external matters, other members of the caste have gradually transferred to the "politicians" that part of the head man's role-set which had to do with problem solving within the caste. An omnicompetency is thus imputed to them, whether or not they merit it.

Today's head men still exercise some authority and continue to officiate at ceremonies in their own *thoks* or neighborhoods, but they are mere shadows of their former selves. On one observed occasion, there was an argument between four young men and a head man who had reprimanded them for gambling outside the neighborhood on the main road. One young man replied,

gambling is a little thing for the improvement of the community [that is, to harm it]. What ought to be stopped is drinking and allowing young girls to go to the market and sell tacks, or women to buy meat in the market. All these things should be stopped; if all would cooperate, then there would be improvement. To pick on just gambling and allow all other bad habits has no meaning especially when head men do it themselves.

This is but an example of an attitude that is growing and is due to the increased politicization and increased integration of Jatavs into other institutions such as the courts and schools.

What is happening, then, is that the hereditary caste head men and "big men" are being supplanted by the "politicians". The functions of leadership, problem-solving, and adjudication of disputes have been differentiated out of the caste and integrated into the institutions of the city and the nation, into party politics, the legislatures, the courts, and the whole structure of patronage and influence involved in the developmental programs of the modern government. In other words, through dichotomization the status of head man was first redefined as "big man" and later was redefined as "politician" and has been integrated into political institutions such as parties and legislatures.

While the leadership function of the head man in a *thok* is changing, *thoks* do nevertheless continue to perform with vitality as units of self help, cooperation, and commensality. These functions are most evident when a marriage of a boy or girl from the *thok* is to take place, although they are evident at other times, too. At the time of a marriage, for example, not only do *thok* brothers assist in preparing and serving the feast, and helping with other odds and ends, but they also contribute financially through the custom of *nyota*. Before the marriage takes place, the father of the girl or boy collects *nyota* or a sum of money from every household in his *thok*. Strict accounts are kept by both giver and receiver and the head man acts as a public witness of each transaction for the entire *thok*. The money given is really a loan because it must be repaid when the original giver has need to ask for *nyota* for some ceremony or other. In this sense, a *thok* is a circulating credit association.

The manifest function of *nyota* is to allow for the accumulation of a large amount of money at a specific time without going into debt to a money lender. *Nyota* also functions latently as an integrative mechanism defining the social boundaries of the *thok*, and it keeps its members attached to it. Members of a *thok* are obliged to give and receive *nyota* only among themselves; outside of the *thok* it is by individual choice. In return for *nyota* collected on a particular occasion, the collector is expected to give a feast called *jyonar* or *mara* in which commensality symbolically unites all members of the *thok* into a mutually dependent group of peers. Unfortunately, there is also an element of prestige involved. A rich man often gives a great feast at a marriage, or at a death feast of a deceased parent. Moral pressure is put upon others less well endowed financially to do likewise in a display of conspicuous consumption. Thus, the money collected in *nyota* is sometimes entirely spent upon the feast, and its original purpose is defeated when the poor man finds himself able to walk with honor

but doubled over under the burden of debt. As one man put it
to me, "One cannot talk with honor among his brothers unless he
does so." (*"Biradari men batcit karne men izzat nahin hai."*)

Thoks also function as units of affective relationships. Each
thok has a *panchayat* meeting place, usually an open space of some
sort wherein the *panchayat* meets. Here too men congregate to
talk, smoke, and sleep in the summer; children play; and feasts
and public entertainments are held. During the rainy season various types of entertainment may be given here such as comic
opera (*ras*); the singing of epic verse (*alha*) about culture heroes
such as Alha and Dr. Ambedkar; and the singing of devotional
hymns (*satsang*). Members of other *thoks* may come and watch, but
must do so as individuals, not as members of the *thok*. Women of
the *thok* are expected to confine themselves to the area of the
thok and to their immediately contiguous neighbors. They will
go out of the *thok* usually on four legitimate occasions: to get the
daily water; to go to a hospital or doctor; to visit relatives; and
to go to the latrine. Men are expected to do the marketing. The
thok is, therefore, an affective unit bounding the spatial mobility
and, to some extent, the face-to-face social interactions of women.

Neighborhoods

In some neighborhoods (*mahallas*) in the city there is only one
thok, and therefore the social organization of the *thok* and the
neighborhood is identical. But in other neighborhoods, as in
Bhim Nagar, *thoks* are only component parts of the total neighborhood. What, then, is the neighborhood as a unit of social organization?

As we have already noted, a neighborhood is a residential ward
in the city. Outside of a neighborhood, people are identified as
so-and-so of Bhim Nagar or as such-and-such of X neighborhood,
and an individual feels a definite identification with his own
neighborhood. Within a neighborhood, however, the word "neighborhood" is relevant to context and may also be used to refer to a
person's particular *thok*.

From time to time a neighborhood can act as a unit. On one occasion a boy from Bhim Nagar was working in another neighborhood. He got into an argument with some people of that place, and in the heat of the controversy he was beaten by them. Another man from Bhim Nagar also happened to be there at the time, and he immediately came to the defense of the boy, even before looking into the merits of the case. He took action because the boy belonged to his own brotherhood (*bhaiband*) of the neighborhood and blood, even the blood of a fictive kinsman, runs thicker than water. Thus, he too was beaten. Another instance occurred during the First All-India Buddhist Conference in Agra, 1963. Rather than contribute to a general fund being taken up by the All-Agra Committee, the people of Bhim Nagar decided to run a kitchen and supply free meals to those from outside of Agra, as a mark of hospitality. They said that Agra's reputation for hospitality was good, and it was up to Bhim Nagar to prove it.

As residential units neighborhoods play an important function in kinship and marriage. All the people of a neighborhood are related, either as real or fictive kin. All such relatives are called a *bhaibandh*. Kinship terms are generally applied to all in the neighborhood. Any widow, divorcee, or unmarried girl is thus a daughter of the neighborhood and is therefore under the protection of its men. Those in the parental generation are called *caca, caci* (paternal uncle, aunt); and those in the grand-parental generation are called *dada, dadi* (paternal grandfather, grandmother). Children, real and fictive, are generally addressed as *mora, mori* (male child, female child), while peers are generally addressed by their own name. Because all in a neighborhood are kinsmen, the neighborhood functions as an exogamous unit; the exogamous interdict extends to the immediately adjacent neighborhoods also. However, it is possible for two children from the same neighborhood to be wed under the specific condition of not having been born there. The marriage of two children born in other neighborhoods who subsequently moved to Bhim Nagar with their parents is considered valid but is not a favored alternative.

As already mentioned, an individual becomes a member of the neighborhood either by birth or by application for membership in a *thok*. The defining criterion of membership is whether or not one has the right to ask for *nyota*. Those who live in the neighborhood but who are not members of any *thok* are called renters (*kirayedars*), although a *thok* member may also live in a rented room or house. Renters, who are better considered as outsiders, are always spoken of with some disparagement and are somewhat looked down upon by the householders (*gharwalas*) or members of a thok (*thokwalas*). Just as one retains an ancestral affiliation to a village, so too does the Jatav retain an affiliation to his ancestral household and its neighborhood. As Cohn (1961) has pointed out, in the village the Camar's tie was to his patron (*jajman*), not to land. Being for the most part landless in the village, the Jatavs easily became an urban proletariat, because they had no roots to anchor them to the rural soil. In the city, however, roots have been struck, because many have been able to buy their own land and homes and have become an urban yeomanry.

In addition to the *panchayat,* which each *thok* had and has, there was also a *panchayat* for the whole neighborhood. This *panchayat* attempted to solve inter-*thok* disputes, cases serious enough for a larger body to consider, and matters of importance to the whole neighborhood. In pre-independence days, each neighborhood was linked to other neighborhoods in and around the city through the *panchayat* system. Each neighborhood was in an intermediate level of *panchayats* called the fifty-three, twelve, and eighteen neighborhood *panchayats* (*trepan, barah,* and *atharah mandi panchayats*). These were again all united into the highest level *panchayat,* which covered the whole city of Agra, called the eighty-four neighborhood *panchayat*[9] (*caurasi mandi panchayat*). These two levels acted like a Court of Appeals and a Supreme Court. The seriousness of the case dictated the level of the

---

[9] This name indicates only the actual number of neighborhoods in the *panchayat* when first established; with further addition of members, the original name remained.

Women making a few
cents by selling fruit and
vegetables, and (right)
Jatav boy and girl
leaning on support for
newly installed light
post.

panchayat to be called. Each of these higher level *panchayats* also had a leader (*sarpanch*) and a mace bearer whose positions were hereditary.

In these higher level *panchayats* the status of *panch* (member of a *panchayat*) was an achieved one insofar as one had to distinguish himself in oratory, wisdom, and astuteness in getting to real issues and finding workable solutions. These higher level *panchayats* made decisions for that part of the caste under its jurisdiction. In addition to their leadership and problem-solving functions, these higher level *panchayats* functioned as outlets for personal ambition and as symbols of the unity of the caste. They also were integrating mechanisms and communications centers for the Jatavs, because in them news was passed on, contacts were made, and friendships were renewed.

It is the *panchayat* system of the Jatavs which seems to have suffered most from the changes taking place in urban Agra. All of my informants date the disintegration of the *panchayat* system from the introduction of parliamentary democracy and politics, although the process had probably begun earlier. While this change was fundamental, there were other factors which also seem important to me. These were: increased wealth through involvement in the market system; increased education; the use of the courts as an alternative system of justice; and the politicization of caste.

Some of the effects of increased wealth have already been discussed in Chapter III. One further effect was that it allowed or increased the means to corrupt. Traditionally the disputants in a case gave a payment to the *panches* for convening the *panchayat*. This sum was spent on food and drink. Increased wealth, however, helped change these traditional payments into secret bribes. Thus, faith in the justice of the *panchayat* system has died.

Increased education has weakened the *panchayat* system and the sanctions which gave it the strength to enforce its decisions. In this sphere Bhim Nagar's achievements appear typical, and it is worthwhile to take a more detailed look into them. Table A–IV of the Appendix shows the amount of education present within Bhim Nagar. My estimate is that about five years of education are needed to leave a lasting literacy. By this criterion there are 145 literate individuals, of whom 130 are male and 15 are female, within Bhim Nagar.

Of the literate males, one boy has a second class B.A., another has an M.A., and three are now studying for the B.A. Three others have received or are receiving diplomas from the Institute of Technology, which will give them specific qualifications for certain technical jobs. Bhim Nagar has also produced a graduate engineer who holds a high post with the railroad and another B.A. who is a deputy collector. Neither of these men, however, were counted as residents in the neighborhood. This record for education within one neighborhood is encouraging when one notes that the first B.A. for the whole Jatav caste in Agra City was awarded in 1926.

How has this educational leap forward affected the *panchayat* system of the Jatavs? First, the educated man, or at least the literate man, is more adapted to the new social environment in which he exists. No longer is there the same respect for the elderly man and his experiential wisdom; these particular criteria for leadership and authority have been partially replaced by a high regard for education and the advantage it gives in dealing with the new socio-political order. The educated themselves have less fear of using the courts, since they have some measure of control over the written procedures used therein. They also feel that it is a matter of prestige for them to use the courts rather than the traditional *panchayats*. Thus, education gives the educated, or literate, man an advantage over the illiterate man when engaging in "bridge actions" between the old system of *panchayats* and the new system of the courts.

Second, education in the broadest sense has meant the communication of new concepts. Urban Jatavs have been exposed to new egalitarian and individualistic ideas through the independence movement, through the teachings of Ambedkar and Gandhi, through the schools for the young, and, most importantly, through party politics for all. Thus, there has been an undermining of the old ideas of caste and traditionalism by the new ideas of democracy and socialism. Since the Jatavs, under the influence of the Buddhist movement, are anti-Hindu, there is also an undermining of the religious belief that the *panchayat* speaks as the word of God and has the power of supernatural sanctions. These new ideas are institutionalized in the courts and politics, and it is through them, rather than through the traditional *panchayat,* that Jatavs are coming to relate themselves to others in cases of conflict and so forth.

Another important reason for the decline of the *panchayat* system is that the courts now furnish an alternative system of justice. Thus, there is the opportunity to engage in "bridge actions" and either to manipulate the *panchayat* or to ignore it altogether.

While I have definite evidence of use of the courts in pre-independence India for certain inter-caste conflicts, I was unable to trace any intra-caste court cases, although no doubt they existed. Even so, it is probable that Jatavs used the courts reluctantly in pre-independence India, because there were few elected Jatav officials who could intervene for them and because, as has been pointed out by Galanter:

In the reported cases it can be observed that the local officials were almost uniformly unsympathetic to the claims of the lower castes—even where these claims were subsequently vindicated by the High Court. Though it is impossible to say how typical were these lower court proceedings, they suggest that at the level of the magistrate's courts, assertion of rights by groups suffering customary exclusion or disabilities met with little encouragement if not active hostility. . . . It should be emphasized, however, that these prescriptive rights and disabilities received their greatest governmental support not from direct judicial enforcement but from the recognition of caste autonomy—i.e., from the refusal of the courts to interfere with the right of caste groups to apply sanctions against those who defied these usages (Galanter 1963: 548).

There was also probable reluctance to use the courts because of lack of education and knowledge of court proceedings.

Under the new Constitution, castes in India still retain their right to discipline their own members, but their ability to do so has weakened because of the increased possibility of "bridge actions." Much of this is due to education, as I have mentioned above. There is now less fear of the courts, since the total external social environment of Indian society has redefined the relationship of the Jatavs to the courts more in terms of the status of citizen who has a right to use the courts than in terms of Untouchable who has learned to fear the courts. "One of the basic themes of the Constitution is to eliminate caste as a differential in the relationship of government to the individuals—as subject, voter, or employee" (Galanter 1963: 555). Since the courts need not judge the real issue at hand, and cases are often trumped up, one party can use the courts as a means to get even with or to wear down another party. The threat of such action is sometimes enough to keep a poor man from starting a case. In effect, the court is like a counter-sanction of greater force than that of the *panchayat,* since it can inflict jailing and fines.

While it is the rich who can most easily resort to the courts and use them as a threat against the poor, recourse to the courts is by no means confined to the rich. One who is dissatisfied with the decision of the neighborhood *panchayat* may prefer to bring the case to the courts rather than to the upper level *panchayats.* Courts are used not only to obtain a just decision; they are also a means to score a point or win an argument. Since members of a *panchayat* often have good knowledge not only of the issues but also of the personalities and reputations of the parties involved, it is often more advantageous to take a case to court where the issues are less known and where there are various informal mechanisms for delaying or influencing a decision. As Cohn (1959) has pointed out, the courts operate on an entirely different set of assumptions than the traditional Indian adjudicative process. The assumptions of the courts are (*a*) equality of the parties; (*b*) no compromise, only a right or wrong decision; (*c*) only one issue in one case; and (*d*) relations based on contract, not status. Traditional *panchayats* assume exactly the opposite.

The third way in which the courts have undermined the Jatav *panchayat* system and its sanctions is that the courts also stand as a reference group from which new standards in the administration of justice are learned. This is evident in the statement of one reliable informant, who said of the *panchayats:* "They have become just like the courts. You pay the *peshgar* [the clerk who controls the order of presentation or docket of cases in the court] a little and he gets your case fixed on time. You pay your witnesses and you win your case."

The final factor which has influenced the disintegration of the *panchayat* system of the Jatavs is the "politicization" of the caste. I have already discussed this to some extent in dealing with the declining position of the head men. When matters of concern or decision for the whole Jatav caste have to be decided, the "politicians," not the traditional *panches* and head men, do it today. Occasionally the "politicians" are even called in to assist a neighborhood *panchayat* when expert advice is needed. Parenthetically, I might add that a meeting of Agra Republican Party leaders resembles in its procedure a traditional *panchayat*. What is changed, however, is first, the content of its discussions and deliberations, which nowadays are largely of a political nature; and second, the method of recruitment, which today is one of personal achievement.

The effect of "politicization" is not confined to leadership but has also penetrated the structure of the *panchayats* and factions within the neighborhood itself. Political contacts have been used to influence the police in favoring one party or another in intra-neighborhood conflicts. The same is true for the courts, wherein political contacts are used to hasten, delay, or influence a decision. These are power factors over which the *panchayat* has little control. Even in the neighborhood *panchayat* a case will generally not be heard once it has reached the courts or until the parties agree to discontinue the case in the courts.

The combined effect of these changes on the *panchayat* system has been differential. The 84 neighborhood, or all Agra *Panchayat*, is for all purposes dead. The middle level *panchayats* continue to meet occasionally, but in a perfunctory way. Only the neighborhood and *thok panchayats* exist with some vitality, but they are mainly deliberative. Even these *panchayats* have little power to enforce their decisions or sanctions. Bhim Nagar's recent experiences clearly point this out. A few days after the holiday of Holi there used to be an annual *panchayat* in Bhim Nagar. In it the rules of the largest and oldest *thok* were discussed, and unfinished business for the year was brought to a close. In 1962, at such a meeting, a set of 24 rules was drawn up concerning what should have been proper behavior for members of the *thok*. These rules reflected new and old problems with which the *thok* was concerned. These included, for example, forbidding attempts to bribe the *panches;* setting an age of marriage in accordance with the new laws; forbidding drinking in marriage parties; and forbidding gambling within the neighborhood. When the annual *panchayat* was assembled again in 1963, after much delay and with half-hearted participation, it was agreed that all the new rules of 1962 had been broken and that nobody had observed them. An abbreviated list of 13 rules, dealing mostly with the timing of feasts, the amounts of *nyota,* and gambling within the neighborhood, was once more drawn up, and all agreed to it. In 1964, however, it was impossible to get the annual *panchayat* assembled once again. It was considered a waste of time to meet and draw up rules which nobody observed or enforced. Thus, the *thok panchayats* now assemble only for *ad hoc* cases, usually when both parties agree beforehand to abide by their decisions, although even this is not a certainty.

In addition to the weakening of the *panchayat* system, other factors have also tended to weaken the internal structure of the neighborhood. I have already noted how the *thoks* are units of affective relationships. In the past, men gathered and smoked the water pipe (*hookah*) at the *panchayat* meeting place. The water pipe was a symbol as well as a mechanism of social integration within the *thok*. However, as one informant put it, "Before people began to smoke cigarettes (*bidis*), the water pipe kept everybody together to talk. Now one lights up a cigarette and walks away." While cigarettes have largely displaced the water pipe, what has also happened is that alternative sources of entertainment and affiliation have become available. There are the ubiquitous movies which most of the young men are fond of and frequently see. Jatavs too attend the fairs (*melas*), which add color and excitement to city life. Shoe factories in other neighborhoods are not only places of work but also are centers of interaction and communication. Side by side with the factories are the markets where one Jatav may meet another to gossip, argue, and generally pass the time of day. There are also politics and the various political meetings held throughout the year for either the elite or the whole community. Schools, both private and public, draw the children out of the exclusive neighborhood environment and into an environment which widens their horizons to some degree. And last but not least are the new communications media, the radio, the post, and the newspapers, which the Jatavs increasingly use. All of these factors tend to integrate the Jatavs into an urban and national culture in such a way as to weaken the internal unity and structural control of the neighborhoods and *thoks*.

One more important force for integration and change needs to be mentioned; this is Buddhism. In Bhim Nagar, as in other Jatav neighborhoods, there now stands a Buddhist temple. This shrine, the size of a small room, was built by voluntary contributions of wheat and money by all the people of Bhim Nagar in about 1957–58. The importance of this temple is that it not only symbolizes the unity of Bhim Nagar, but it also symbolizes the integration of the Jatavs into an all-India Neo-Buddhist movement of at least nominally converted ex-Untouchables. This

Left, newly invested head man of a *thok*. Right, a *thok* panchayat meets. *Thokwalas* collecting *nyota* (below) for a marriage.

movement has opened the neighborhood to influences and ideas which are more than purely local. In its attack on caste and the caste system, the movement has indirectly attacked the whole traditional basis of Indian society and has further weakened the unity of the neighborhood and the *panchayat* system. The goal of the Jatavs is no longer to preserve the inner order and integration of their caste and their neighborhoods; rather, they desire to become more integrated into the urban culture of Agra and the national culture of India. The real problems of social disorder and social control, according to the Buddhists, are not those of internal caste cohesion and control but those of the caste system itself.

In structural terms, much of the weakening of internal caste traditions and cohesion is due to the many status dichotomizations occurring among the Jatavs. The status of child (*beta, beti*) has become student in the schools; the status of Camar has become factory owner (*karobar*) or craftsman (*karigar*) in the market system, and again Camar has become plaintiff versus defendant in the courts; the status of Untouchable has become Scheduled Caste in the various institutions set up under the government's "protective discrimination" policy; and the newly assumed status of voter has also become party member in the political party system of India.

These dichotomizations can also be seen as adaptive responses of members of the caste to the potentialities presented in the new environment of independence, parliamentary democracy, and the universal franchise. This environment is to some extent favorable to Untouchables who are now Scheduled Castes and therefore receive special treatment for their betterment under the government's policy of "protective discrimination." The need, therefore, of members of the caste to protect themselves through self-regulation from the dominance of other castes is less. This new environment likewise offers opportunities for greater individual freedom and autonomy from one's caste and its control over one's behavior; the Jatavs are availing themselves of these opportunities.

Summary

It is now possible to summarize how changes in the areas of politics, religion, and economics have affected the social or caste life of the Jatavs among themselves. While there has been a loss of internal caste cohesion and unity, this has been paralleled by an increased sense of self-identity, self-worth, and unity before other castes. Thus, integration and unity vary with the situation. As Buddhists and Republicans, Jatavs are strongly united, and this unity spreads far beyond the borders of Agra City into all of Uttar Pradesh and Maharashtra. In this sense, then, there has certainly been an increase of what Professor Srinivas has called "horizontal solidarity" of caste; that is, the caste boundaries are no longer merely local in extent. Like water into which a pebble has been dropped, Jatav and Buddhist *esprit de corps* has spread rapidly under political, social, economic, and communications changes.

However, we must go a step further than Srinivas, because in post-independence India there has been, for the Jatavs at least, a great increase in the "vertical solidarity" of caste. Caste *panchayats,* head men, *thoks,* and neighborhoods have disintegrated in functional importance. Family life has suffered much less, but even here fathers are now occasionally challenged by their educated sons and younger people have different attitudes toward the marriage ceremony and arrangements as well as toward inter-caste marriages. Yet these disintegrating institutions and attitudes have been paralleled by an increased vertical integration into institutions of the city, state, and nation. The courts, political parties and politics, the politicians, movies, schools, shoe factories and market, and the Neo-Buddhist movement now perform for the Jatavs many of the functions that their own caste once did. In sociological terms, these new secular institutions are functional alternatives to the traditional caste institutions.

The process that mediates between this simultaneous disintegration-integration is dichotomization of statuses. In this process, parts of the traditional role-set of statuses within the caste have been redefined as separate statuses in institutions without the caste. Children are now partially socialized in the schools as students: head men (*chaudhuries*) are now almost figurehead leaders because of the "politicians" (*neta log*), which have by and large supplanted them in the institutions of party politics and parliamentary democracy; Camars have now become citizens, voters, and party members under the universal franchise and party politics; Camars, too, have become plaintiffs and/or defendants in the courts. Dichotomization, then, is the process which links changes external to caste organization with changes internal to it. The concept of dichotomization, moreover, has allowed the analysis of material in this chapter to fit into the analytical framework of role theory used in the rest of the book.

Those who know much about India will recognize from reading this chapter that many of the traditional forms of social organization among Jatavs were similar to those that still exist in North Indian villages. It would seem, then, that up to the recent past there was little difference between the rural and urban forms of social organization in North India and that Indian society was much more homogeneous than the theories of rural and urban sociology would lead us to believe (see Lynch 1967).

 CHAPTER VII

Conclusions

In the preceding chapters a picture of the changing life of the Agra Jatavs has been etched. Throughout the book the theme of descriptive fact has occasionally been counterpointed by theoretical analysis. It remains for us to pull the material together and see how it answers the problems posed in the Introduction. In so doing I should like to discuss—speculate, if you will—what the case of the Jatavs can tell us about the dimensions of social change in India, in particular, and about the relevance of certain anthropological theories for the analysis of social changes, in general. This discussion, then, moves from descriptive conclusions (ethnography) to theoretical ones (ethnology).

Ethnographic Conclusions

In the Introduction, I hypothesized that caste was an adaptive structure, so that given the changing social environment in which castes interact, adaptational changes in caste organization may occur in two ways. First, changes in the internal social organization of caste *tend* to occur in such a way as to preserve caste integrity. That is to say, a caste under conditions of change does not necessarily disintegrate into some other type of social group.

This is the conservative aspect of adaptation. Second, change in the external relations of a caste to other castes *tends* to occur in such a way as to maximize for a caste's population its share of caste external scarce goods such as power, prestige, wealth, and education. This is the creative aspect of adaptation by which new social forms sometimes arise out of old ones.[1] What, then, are the adaptations that have occurred in Jatav caste organization?

In the first place, a caste in the new socio-political environment of India can and often does function as a ready-made association or interest group which can be mobilized for political action and ends. This is what the Republican Party does for the Jatavs. This adaptation of caste by the Jatavs is buttressed by their *en bloc* residential segregation into caste wards and by their population size in Agra City. These have become powerfully adaptive in the new environment of the universal franchise. Jatav votes are now important, not only to themselves but also to members of other castes, who cannot be elected without Jatav support. This was most strikingly evident in the case of the Republican deputy mayor, a man of merchant caste who needed Jatav votes to get elected to the Agra Municipal Corporation. A caste, in this sense, is an organized vote bank which can bargain for the use of its political capital.

The Republican Party is an association whose members include those with traditional caste attitudes and those with modern political attitudes. At present it is led by the Jatav moderates, who are at the same time creative enough to engage in political participation but conservative enough to keep the Party a single-caste Party. To carry the biological analogy of adaptation a step farther, one might say that the moderates are the present phenotypic form of Jatav political adaptation. They balance traditional caste exclusiveness with modern political inclusiveness. In this sense they are hybrids adapted to the contradictory statuses of caste and

---

[1] My thinking here has been stimulated by the book *Evolution and Culture* (Sahlins and Service 1960). Chapter 3, "Adaptation and Stability," has been particularly useful.

citizenship now present in Indian political systems. Yet, within the Jatav population the recessive genotypes of the radicals and the conservatives also exist. The Republican radicals are specialized and adapted to a possible future environment of individual equality and achievement. The Congress conservatives are specialized and adapted to caste inequality and ascription, and for this reason they accept cooperation with the Congress Party. Given a shift to left or right in Indian politics, either of these two groups could surface and gain control of the caste vote bank.

The second adaptive change that has occurred in the Jatav caste is in its leadership. As in the past, the caste is still composed of a small group of leaders and a large group of followers. Yet, effective leadership is no longer in the hands of the caste head men and *panches*. It is now in the hands of the "politicians" and to some extent the "big men," whose positions are achieved, not ascribed. These men now speak for the Jatavs, represent and fight for their interests, and to some extent help solve or advise on problems internal to the caste. This supplantation of the old caste leadership is an adaptation of internal caste organization. On the one hand, it helps to maintain the internal solidarity and unity of the Jatavs. On the other hand, the "politicians" and "big men" help to relate both individuals and the whole caste to new opportunities in the Indian structures of power, opportunity, and wealth. They are "culture brokers," who adapt old structures to new ends.

In pre-independence India, Jatav leadership, and through it the caste as a whole, did grasp the situation correctly in the attempt at social mobility through Sanskritization. At that time Sanskritization was the only legitimate channel open to them. But political tactics were also involved in trying to get the British Raj to recognize Jatav Kshatriya identity and to grant educational concessions as well as appointed political posts. If the pre-independence era can be considered an older era, the case of the Jatavs does confirm Marriott's (1959: 71) hypothesis "that such older instances will prove to involve castes having a broad social or political stance." This is important because it leads to the conclusion that

Sanskritization, when it occurs, is probably only a dominant or overriding technique but not the only technique used by socially mobile castes in India. (See also Srinivas 1966: 94.) If we are to understand Sanskritizing movements in their totality, therefore, we must see them in some larger frame of reference than that included in the concept of Sanskritization itself.

In post-independence India, the socio-political environment has changed and so, too, has the Jatav attempt at mobility. Jatavs have switched from an acceptance of caste through Sanskritic behavior to a rejection of caste through political participation. It is now the "politicians" who are the readers of the fluid social situation and the directors of Jatav action. Political participation through the Republican Party has become the dominant or overriding technique of the Jatavs to get what they want, and Buddhist identity[2] has replaced Sanskritic Kshatriya identity. Previously Jatavs had a manifest goal of higher ritual rank through Sanskritization, while their latent goals were legitimation of the economic success already achieved and legitimate access to upper caste statuses which controlled other strategic resources of power, opportunity, and wealth. Now these latent goals have become manifest because there is little need to seek ritual validation from the Brahmans before legitimate access to strategic resources can be obtained. Jatavs directly seek their ends through the political tactic of pressuring the government and other castes to enforce the principles of equality, secularism, and freedom set down in the Constitution.

The present caste leadership with its three blocs is to some extent pragmatic and open to change. Buddhism and the Republican Party are, in a sense, experiments in adaptation, because they have not as yet proved *wholly* successful for the Jatavs in getting what they want in the structures of power, opportunity, and economy. If the present political stance works, all well and good. If it

---

[2] I have pointed out in the preceding chapters how Buddhist and citizenship status for the Jatavs are virtually identical.

fails, then the Jatavs are aware of other alternatives such as Communism or cooperation with the Congress Party.

The third adaptive change of Jatav caste was to the shoe industry, the market system, and to government policy. The traditional Jatav caste occupation of leather working has evolved into a monopoly of shoe production. Morever, their status as Scheduled Caste can be seen as a government stimulated adaptation of their Untouchable caste status. The government's policy of "protective discrimination" gives some slight advantage to the Scheduled Castes over the non-Scheduled Castes.

There is a paradox in the government's "protective discrimination" policy. On the one hand, it is breaking down caste by opening some channels for individual mobility within the larger society. On the other hand, it creates a vested interest in the preservation of Scheduled Caste status. The benefits of the policy are too real and too immediate for the majority of Jatavs to want to give up the privileges derived from being S.C.'s.

What is occurring is that adaptive changes at the level of caste as a group are being counter-balanced by adaptational changes at the level of individuals. When individuals pursue their personal advantage by using institutions of city, state, and nation such as the courts, the schools, and administrative posts and political offices, they weaken the internal homogeneity, cohesion, and solidarity of their caste. This is because traditional caste functions have been transferred and integrated into modern secular institutions.

Even the advancement of Scheduled Caste individuals under the policy of "protective discrimination" does not necessarily spread its benefits to their caste mates. In pre-independence India an individual's advancement could not receive full validation until his whole caste had Sanskritized. In post-independence India a low caste individual can seek validation and protection in the courts and the principles enshrined in the Constitution, which take the individual and not the caste as the unit of the society.

He has more hope of getting this than he ever had of getting Sanskritic validation. While an individual may still have need of his caste's political support, the validation of his election to office does not depend upon getting other members of his caste to Sanskritize their way of life. This is not to say that caste is irrelevant to the individual in India today. Rather it is to say that the functions caste performs for individuals are becoming weaker and fewer.

In addition to the adaptations I have just mentioned, I have also tried to indicate in the book that much Jatav social behavior in the new environment of parliamentary democracy and the universal franchise did not, like Topsy, "just grow." On the contrary, there are strong continuities with past structure and function, although not in overt form. Buddhism and its modern Bodhisatva, Ambedkar, have strong cultural and ideological continuities with past movements, such as the cult of Kabir. Buddhism is also an acceptable, but revolutionary, ideology, because it is presented to Jatavs in a religious idiom and in a society that positively values religious symbols and expression. Buddhism also continues the theme of "We are the original Indians" found in the earlier movements of the Jatav Men's Association and the Original Hindu Movement, and in one form or another in most caste mobility movements in India. It is an idiom of transition using familiar cultural forms to mediate between traditional caste values and secular democratic ones. The Republican Party, too, is built upon older foundations; it is the successor of the Scheduled Caste Federation and the Jatav Men's Association.

Discontinuity with the past is also present, since through dichotomization Jatavs are being integrated into more inclusive institutions and structures of the state and nation. The courts, schools, political institutions, and the market system have brought about a disintegration of the caste *panchayats* and the traditional head men. The Jatavs have rejected caste, the caste system, and Hinduism. At the same time, they are attempting to replace them with Buddhism and an egalitarian society based on individual achievement and equal opportunity for all. More importantly, the Jatavs are continuing to expand their status-sets so that more and more of them are in statuses cross-cutting those of other castes. Thus, some of the mutually exclusive status-sets of the upper castes are breaking down. This is important, since it creates the minimal conditions for the formation of associations and interest groups on the basis of statuses other than caste. It also tends to weaken the strength of powerful caste groups in the Indian power structures of today.

In the discussion of politics in Agra, I have at times pointed out how "bridge actions" occur when people activate either their caste or their citizenship status, according to what is most advantageous to them. I have also pointed out that discrimination on a caste basis continues to occur; for example, in the failure of the Jatav candidate to be elected deputy mayor. However, given the changing socio-political dimensions of Indian society, this does not in itself seem a regressive step. On the contrary, the increase of "bridge actions" appears to me a sign of advance since it indicates that people are coming to grips with competing loyalties and affiliations and are trying to create in India a "civil politics of primordial compromise" (Geertz 1963: 157). The very process of modernization itself brings forth and exacerbates the competing loyalties of citizenship and caste statuses in the struggle of a new state to become a nation. Thus the problem, at least at present, is not to "wish them out of existence by belittling them or even denying their reality, but [it is] to domesticate them" (Geertz 1963: 128). In this sense, too, caste is an "adaptive structure," since, in the political arena especially, it has become the meeting place of old traditions and institutions with new ideas and structures. Thus, my assessment of the Republican Party and the Buddhist movement concurs with the judgment of the Rudolphs on caste associations in India.

Within the new context of political democracy, caste remains a central element of Indian society, even while adapting itself to the values and methods of democratic politics. Indeed, it has become one of the chief means by which the Indian mass electorate has been attached to the processes of democratic politics (Rudolph and Rudolph 1960: 5).

Political participation has an important leveling function for all castes in Indian society. Upper castes in an attempt to retain their traditional prerogatives have taken to political participation, and by the very fact of that participation they are forced to change. What was traditionally theirs by right, and rite, they must now bargain for with other castes including Untouchables, at least in the political arena. Thus, they have changed in politics their mode of acting and relating to members of other castes. They, too, have had to adapt to parliamentary democracy, because it is a matter of politics or perish. One indication of the changes taking place is that Untouchability in India, like discrimination in the United States, has become a pejorative term; one must avoid publicly practising either.

I must at this point, however, inject a word of caution and restraint. The case of the Jatavs (and the Mahars) is not typical at the moment. While Untouchables seem to be better off and politically more active in the cities of India, the condition of Untouchables in the villages is still unfavorable. In the rural areas they do not seem to have reached the kind of political awareness and participation that the Jatavs and Mahars have. In the villages there are only a few indications that political participation has begun to disturb a centuries-old way of life (Cohn 1954, 1955; Rowe 1960; Beteille 1965b: 28–31; Mahar 1958, 1960).

The paradox for the Jatavs in the present socio-political system of mixed traditional and modern elements is that the very conditions which they are trying to create are also to some extent destroying their present adaptive niche. The most striking example of this is in the shoe market. When the Jatavs argue and fight for the abolition of caste and caste-ascribed statuses they are, in effect, arguing for the elimination of their own traditionally ascribed status of shoe maker. Until the advent of the Punjabis, this status was a profitable monoply for the Jatavs. Now, however, the Punjabis are occupying the status of shoe factory owner and breaking down the Jatav monopoly at least as far as ownership of factories goes. With more and more statuses open to competition and achievement, the Jatavs are at a disadvantage compared to other castes who already have command over the scarce resources of power, money, and prestige. The same changes which are to some extent advantageous to the Jatavs are also advantageous to members of other castes, who enter the new socio-political environment already equipped with greater potential to dominate more and more statuses in the power and opportunity structures, at least in the short run.[3]

In the long run, however, the implication of these changes is that the insularity and separateness of lower castes from upper castes and from integration into all aspects of the Indian society is to some extent breaking down. For example, the courts now are an alternative system of justice, and politicians form an alternative leadership (to hereditary head men) for the Jatavs as well as for other castes. These new adaptations, therefore, are functional differentiations out of a multi-functioned caste into a milli-functioned state and nation. As the functions of a caste become fewer, the strength of an individual's ties to it is weakened. At the same time, the integration of Indian castes and their members into a more egalitarian society and a single Indian identity is thereby advanced.

Ethnological Conclusions

The first ethnological question asked in the Introduction was: Can Sanskritization be defined structurally and can it be fitted into a larger frame of reference that is both synchronic and diachronic? In answer I have demonstrated how the use of reference group and status-role theory enabled me to make a structural

---

[3] I hesitate to pass judgment on the Jatavs in the form of Eric Hoffer's comments on people like them. I hesitate because I feel that the Jatavs have not yet had full enough taste of freedom to know whether or not they really want to live with it and all the responsibilities it brings. I leave it to the reader to make his own decision. Hoffer (1966: 37) says, "Those who see their lives as spoiled and wasted crave equality and fraternity more than they do freedom. If they clamor for freedom it is but freedom to establish equality and uniformity."

definition of Sanskritization. In so doing I have concurrently operated on two levels, the concrete and the analytic.

At the concrete level a description of the situation, in which the Jatavs engaged in mobility-directed action, was given in terms of how the Jatavs defined it. In order to Sanskritize, they claimed to be Kshatriyas and attempted to validate this by imitating the Arya Samaj. In trying to assert their claim the Jatavs saw themselves blocked or in conflict with orthodox Brahmans and other upper castes who considered them Untouchable Camars and not Kshatriyas.

At the analytical level, however, a definition of the situation for a mobile group was specified in terms of three generalized types of reference groups; that is, identification, imitation, and negation. Actual interactions, which followed such a definition of the situation, were analyzed in terms of generalized types of statuses which allowed for synchronic and diachronic comparisons. Thus, Jatavs claimed dominant status of Kshatriya but were rejected by others who activated their salient status of Untouchable. Such interaction more often than not resulted in conflict when Sanskritizing or contradiction when engaging in political participation. Jatavs also make use of the controlling status of Hindu in order to occupy the structurally linked status of Scheduled Castes which gives them access to the opportunities granted under the government's policy of "protective discrimination."

The point that I wish to make here is that the use of these theories of reference groups and status-role enabled me to make a structural definition of Sanskritization not in terms of cultural attributes and symbols, but rather in terms of social structural positions and relations. The identification of these structural positions and relations in terms of a generalized set of concepts, such as reference groups and dominant and salient statuses, has enabled me to get out of a major difficulty with the Sanskritization concept; namely, *it is culture bound*. These generalized concepts provide a theoretical frame of reference with which mobility movements in India can be compared to mobility movements in other countries.[4] This frame of reference has both synchronic and diachronic applicability. There is one further advantage. The use of such a generalized frame of reference has helped me to side-step the proliferation of terms which is now arising on the Sanskritization analogue, such as Kulinization (Prasad 1957), Kshatriyaization and Desanskritization (Srivastava 1963), and Pali-ization (Zelliot 1966), a term I would otherwise have been compelled to use. While these terms tell us something of the cultural uniqueness of the movements they describe, they tell us nothing of the social and structural commonalities underlying them. It is to this latter task that I, as an anthropologist, have addressed myself.

The second of the three problems posed in the Introduction was: Is there a functional alternative to Sanskritization? I have tried to demonstrate how a caste, such as the Jatavs, may give up Sanskritization for political participation, because it is now advantageous for them to do so. Structurally this is due to the fact that the Jatavs now occupy and activate the dominant status of voter. Scheduled Castes, such as the Jatavs, by activating their status as voters have been able to elect their own caste-mates to statuses of some power and prestige within the power structure. Scheduled Caste observability over the role performance of other castes is not only made possible but also sanctioned by the law of the land. Such reverse observability of a low caste upon upper castes is almost impossible within a caste system.

Since the contradictory statuses of caste and citizenship can

---

[4] The analysis of the Black Muslim movement in America by Essien-Udom (1964) in his book *Black Nationalism* can be easily fitted into this frame of reference. The Black Muslims take African Muslims as their reference group of identification. Their reference group of imitation, however, is middle class Americans whose habits and way of life are reflected in the commandments of the Black Muslim religion. Their reference group of negation is the white man, particularly the white man in America. The reference groups of imitation and identification of the Black Muslims help us to understand how they can identify with other Muslims and yet be very unlike them in doctrine and belief. The structural fact on which the movement is predicated is that when the Negro tries to activate his dominant status of citizen in America, his salient status of Negro is activated against him by whites in discriminatory practices. This fact structures the situation which is defined through reference groups.

both be asserted as dominant, there is the possibility of "bridge actions" in actual situations; that is, either caste or citizenship status can be activated depending upon what self-advantage dictates in a particular situation. Contradiction also exists in the structure of Indian society because citizenship as a controlling status and achievement as a recruiting principle vie with caste as a controlling status and caste ascription as a recruitment principle to the structures of opportunity and power.

When Scheduled Caste status is activated vis-à-vis non-Scheduled Castes, there is the possibility of the intervention by a third party in the person of the government, the Commissioner for Scheduled Castes and Tribes, the Harijan Welfare Officer, or the courts. This third party has structural observability over the role performance of other castes and is not neutral between citizenship as a dominant status and caste as a dominant status. In such contradictory circumstances, the law says that citizenship should be dominant, at least formally and publicly.

The analytical concept of observability proved most useful in the discussion of the Jatavs. It led me to ask what were the consequences in terms of information (and to some extent power) gained when they occupied statuses with high observability in the Indian political and administrative structures. The consequences of observability acquired through the occupation of the sinecure statuses granted to but one Jatav in the pre-independence days were not minimal. Much information about political decisions and activity was acquired, and this led to pressure tactics and a revised political strategy on the part of the whole Jatav caste. The same is true of the few MLA's of the Republican Party in Agra today.

I would hazard the prediction that political participation as presented in this book is the path that mobility movements will increasingly follow in India. The danger lies in whether or not the more conscious and vociferous demands of these movements can be met in an economy of scarcity before they jeopardize the goal of democratic socialism that India has set for itself (see Weiner 1963).

I would also hypothesize that the theories of Friedl (1964) and Frazer (1963) can be translated into the theoretical notions of reference group theory and observability. These two theories, "lagging emulation" and "one-way visibility" respectively, attempt to explain the Sanskritization of lower castes and the concurrent Westernization of upper castes in village India through the notion of differential observability of imitative reference groups. The lower castes with little mobility and few urban contacts imitate the local dominant caste,[5] while the dominant castes having greater mobility and urban contacts imitate the urban prestige model. These two theories, when viewed in a framework of reference group theory and the notion of observability, eliminate the need for three separate theories. Thus, what is happening in villages can be analyzed with the same theory used for cities and can be seen as identical in process.

The third of our three Introductory problems was: Can a structural description of the effect of external changes in a caste's relation to other castes upon its internal social organization be given, and can this be fitted into the same frame of reference used for Sanskritization? The answer is yes, because it only requires a fuller exploitation of status-role theory already used in the analysis of Sanskritization. The differentiation of various new statuses out of the caste was identified, and through the process of dichotomization their integration into various caste external institutions of Indian society at city, state, and national levels was noted. In this sense, then, change within the caste was seen as an adaptive response to changes in its external relations to other castes and to changes in the institutions of the Indian social system.

[5] The notion of a dominant caste is entirely different from that of a dominant status. "Numerical strength, economic and political power, ritual status, and western education are the most important elements of dominance. . . . When a caste enjoys all or most of the elements of dominance, it may be said to have decisive dominance." (Srinivas 1959: 15)

Perhaps I have overstated the notion of dichotomization throughout this book. My intent in doing so, however, has been to underline its importance and utility in the analysis of social change, especially in countries like India, where a planned attempt at change, organized at higher levels of government, is being made. Dichotomization links the differentiation of functions and statuses *out of* traditional forms of social organization with their integration *into* new institutions. It forces one to look at both at the same time. In this sense, I disagree with Smelser (1964), who asks that differentiation and integration be considered separately; they are sometimes better considered together when linked through dichotomization. My contention is that in societies with government planned and organized social change, it is the many new government projects, offices, and institutions which are stimulating status differentiation by drawing people out of traditional forms of organization, such as caste, and integrating them into modern institutions at city, state, and national levels. The "politicians" (*neta log*) of the Jatavs did not exist in a vacuum after differentiating out of the status of head man (*chaudhury*). On the contrary, the differentiation of caste leadership statuses was stimulated by the fact that new institutions were opened to Jatav participation. These new institutions required new leaders who were specialized in performing a few roles of the old head men as well as new ones. Thus, dichotomization has led me to look at how more complex forms of social organization, such as the courts, the market, and the schools, are breaking down the simpler ones of a caste, such as the *panchayat*. It has also enabled me to demonstrate how a society based on mutually exclusive status-sets (castes) is in slow transition to one based upon cross-cutting status-sets (classes).

Some further points which have been implicit in the analysis now need to be made explicit. The concept of Sanskritization, without a doubt, has revolutionized our understanding of Indian society. It has helped us to see that the Indian caste system was not completely static and was subject to change. It has also proved a powerful, but culturally restricted, synthetic concept which has drawn together much data and many problems in such a way that they now make some sense. However, it seems to me that we have been so dazzled by the new frontiers of understanding which the concept has generated that we have failed to go beyond them or even to look on either side of them. Sanskritization has become so completely self-explanatory that once mentioned there is little need to look for other processes that accompany and modify it. Yet we have seen that the case of the Jatavs seems to confirm Marriott's hypothesis that Sanskritization is probably accompanied by other techniques of mobility and that its manifest function of a rise in ritual rank is also accompanied by the latent function of gaining legitimate access to strategic resources. Sanskritization, therefore, now needs to be examined in concrete cases to see what are the manifold implications and sub-processes that accompany it. It also needs to be examined within the context of other more general theories of social change. Indeed, Srinivas's own plea for studies of local and regional Sanskritization has been little heeded by those of us who have followed in his footsteps. The major problem with the concept is that it is a heterogeneous concept.[6] It includes social and historical processes, cultural items and processes, assimilation and acculturation of tribals, processes of mobility within the Indian caste system, and various other items of patterned behavior and belief. It is precisely here that field studies, hard thinking, and theoretical construction need to be done.

Among the many meanings of the term, I have restricted myself to that most generally used. In this sense Sanskritization refers to a technique for social mobility (see Marriott 1959: 71); that is, it is considered a means, a way, or a strategy for getting up in the caste hierarchy in the same way that getting (keeping) up with the Joneses conceptualizes the technique for getting up in the Ameri-

[6] "It is necessary to underline the fact that Sanskritization is an extremely complex and heterogeneous concept. It is even possible that it would be more profitable to treat it as a bundle of concepts than as a single concept. The important thing to remember is that it is only a name for a widespread social and cultural process, and our main task is to understand the nature of these processes. The moment it is discovered that the term is more a hindrance than a help in analysis, it should be discarded quickly and without regret" (Srinivas 1962: 61).

can class system. But Sanskritization is often used along with other techniques, such as political pressure. This leads to two points. First, it would seem preferable to consider Sanskritization the overriding, but not the exclusive, technique for rising within the Indian caste system. Therefore, other processes and techniques also ought to be considered in further analyses. The second point is that Sanskritization is of too low an order of generality and is too culture-bound to be useful for purposes of comparison and contrast. I suggest that it, and getting up with the Joneses, be subsumed under the general process of elite emulation. Elite emulation is sufficiently generalized and culture-free to enable us to make comparisons and contrasts with other generalized techniques such as revolution, passing, and political participation.[7] It also gives us a concept under which cross-cultural comparisons can be made. This Sanskritization does not do, since it occurs only in India. The job of the anthropologist not only is to describe mobility movements in all their cultural wholeness and uniqueness, but also it is to derive general social processes which occur across cultures. Sanskritization looks to the former problem; elite emulation looks to the latter.

This, then, brings me to another point which should be made explicit. When we consider Sanskritization in India, we look at it as a process of mobility within the Indian caste system. It seems to me that we sometimes are led off the track by the word mobility, because it connotes moving *up* in a hierarchy. Thus, the goal of Sanskritization, and mobility movements elsewhere, is often seen as getting up in the hierarchy and this becomes "the end" in itself of the movement. I suggest that we consider the question from a slightly different point of view and shift from the focus of moving *up* to the focus of moving *into* statuses granting access to strategic resources. From this point of view new questions arise. Why do the X want to move into a higher rank status? If higher rank status is viewed as a controlling status, then what other statuses are entailed in, or legitimately combinable with, it? What access to

---

[7] I am indebted to Dr. Herbert Passin for pointing this out to me.

strategic resources such as wealth, power, and education does occupation of these other statuses give the X? The Jatavs wanted to become Kshatriyas not just because it gave them higher ritual rank; that was only the manifest function of their actions. The fact is that they wanted higher ritual rank, because it also promised them legitimation of new statuses already occupied and legitimate access to others not yet occupied. And, these new statuses gave the Jatavs access to the strategic resources they wanted.

I would suggest that the preceding statements also answer Berreman's (1966: 292) "crucial question [which] is, upon what bases does a group choose reference groups?" Reference groups are chosen in a way that strategically defines a group's situation before other groups with whom it must come into contact as it seeks to change its position in the society by mobility-oriented behavior. The definition of the situation in terms of reference groups is, as it were, a strategy or plan for gaining and legitimizing access to strategic resources in a particular society.

The analytical concepts and procedures used in the discussion of the Jatavs and of Sanskritization have an obvious relationship to what Martin Orans (1965) in his book on the Indian tribe, the Santals, has called the rank concession syndrome (RCS) and the emulation-solidarity conflict (ESC). The RCS, if I understand it correctly, means the concession of superiority to another's group and the consequent acceptance of the inferiority of one's own group on all dimensions of ranking. This is not the same as power concession, since power concession merely involves acknowledging that the superior group has more power or is higher on the dimension of power, but is not culturally superior or superior on all dimensions to one's own group.[8] In the RCS the acceptance of one's own inferiority leads to cultural and social emulation of the superior group in order to rise in rank by becoming identical or similar to it. When power alone is conceded, it merely leads to emulating the techniques of power usurpation and power main-

---

[8] For a discussion of the problems of multi-dimensional ranking, especially as it applies to India, see Berreman (1965).

tenance of the more powerful group. In the RCS social and cultural emulation, according to Orans, leads to the ESC. In it, a group which is emulating another suffers a loss of its own internal solidarity and cultural integrity in proportion to the amount of its emulative activity. A group which emphasizes its own internal solidarity and integrity before other groups decreases in a like proportion its emulative activity. Thus, there is a conflict between the exogenous pull of emulation and the endogenous pull of solidarity. A possible resolution of this conflict is through the development of "naturalization"; that is, the incorporation of items from the other's culture and reinterpreting them as one's own. Another possible solution is through the development of one's own Great Tradition, "which will simultaneously preserve distinctiveness and solidarity on the one hand, and express their [the Santal elite's] own deep evaluations of what is true, beautiful, and worthy of rank on the other . . ." (Orans 1965: 132).

Among the paths to higher rank, according to Orans, the most important are the economic and the political. The economic path is that taken by individuals to improve their economic position so that they can gain the means to adopt the culturally relevant symbols and activities of those in higher prestige rank. This path emphasizes individual emulation and success at the expense of group solidarity and progress. The political path on the other hand is a movement to improve the social and political (and, I believe, ultimately the economic) position of all members of the group in their society. Political movements tend to develop a new Great Tradition in order to resolve the ESC.

The study of the Jatavs presented in this book leads me to make a few qualifications on the RSC and ESC theories of Orans. First, the path of economic mobility for the Jatavs did not decrease solidarity, it increased it. This was because the "big men" who had prospered economically could not raise their caste rank without pulling their caste mates along with them. Thus, they started the Jatav Men's Association and tried to create a new sense of caste identity, caste solidarity, and caste worth. It seems to me, then, that the economic path to higher rank is likely to decrease solidarity only in societies where ranking recognizes individual achievement. It does not necessarily decrease solidarity, at least when higher rank is the goal, in a society organized through a caste system which ranks groups and not individuals.

A second qualification, which seems implicit, if not explicit, in Orans's theory is that the RCS does not seem to be operative in those groups which have adopted an alternative Great Tradition, such as Buddhism for the Jatavs and The Sacred Grove Religious Organization (Sarna Dharam Samlet) for the Santals. Neither of these two groups concede rank to the Hindus. On the contrary, they vaunt their equality, if not superiority, before the Hindu upper castes. The only thing they do concede is power (Orans's power concession) and in so doing they emulate only the power-usurping and power-maintaining techniques of the upper castes. The Jatavs have their Republican Party and the Santals, during the time of Orans's study, had their Jharkhand Party to deal with this problem.

The theory of reference groups makes this clear. A group may select different reference groups for different purposes. A group's reference group of identification (solidarity) with all its cultural elaboration can be distinct from its reference group of imitation (emulation) with its concession to superior power only, as Orans perceptively noted. Thus, even such a highly emulating group as the Jatavs need not suffer any loss of solidarity, nor need it ever suffer an emulation-solidarity conflict. The solution lies in anchoring these two activities on two different concrete reference groups.

A final qualification to Orans's theory involves a point I have made many times but which bears repetition. The RCS theory has been developed on the basis of two groups or actors in the situation and does not consider what happens when a third party that can mediate between them enters. Orans notes (1965: 128) a political-judicial encroachment of Hindu courts and panchayats upon the Santal, which tends to break up their internal solidarity.

But does it do so to Santals alone? In post-independence India, the modernizing nation with its democratic constitution, Five Year Plans, and parliamentary politics works on both upper Hindu castes and the Santal (and Jatavs) to break down their internal solidarity and integrate them both into an emerging social system. The many contradictions that exist in Indian society among caste, tribal, and national forms of social organization are evidence to me that India is a state in transition to becoming a nation. Upper caste Hindus, lower caste Buddhists, and tribal Santals are being forced to make concessions to the third party, the nation, which mediates between them. This does not seem to be a simple rank-concession situation. It is one in which the third party enters, mediates, and unites the two into a new identity, although the day when this is fully achieved may be far off. The third party, the government, is the structural reason underlying the new solidarity not only of "all the Santal with all other tribals of the area" (Orans 1965: 119), but also of the Santals and other tribals with Indian castes when they enter politics, the administration, and other institutions of developing India.

I wish to conclude by saying that my analysis of Sanskritization does not mean that I advocate abandoning the term. It has been and will continue to be useful, at least in India. My intent has been threefold. First, I have attempted to show the other side of the Sanskritic coin; that is, the social-structural side as opposed to the cultural side. Second, I have attempted to provide a frame of reference in which mobility movements in India can be compared to mobility movements outside of India. And third, I have tried to illustrate through the case of the Jatavs some of the fundamental non-Sanskritic changes that have been taking place in India since independence.

GLOSSARY

The following conventions are used in this glossary: Double vowels are for long vowels. A sub-script dot (e.g., ṭ) is for a retroflex consonant. Aspirates are followed by *h*. Velar *n* is marked by a superposed dot (e.g., ṅ).

I. WORDS

Transliteration	*Hindi*	*Gloss*
alha	aalhaa	epic verse
alh-khand	aalhaa-khaṇḍ	the epic of Alha
andolan	aandolan	campaign, upheaval
arhati	arḥatii	broker, factor
atharah	aṭhaarah	eighteen
avatar	avataara	incarnation (as of a god)
bakal	bakal	entrance and courtyard around which are houses
bandhan	bandhan	a tying, something bound about
baniya	baniyaa	merchant
barah	baarah	twelve
basti	bastii	neighborhood, ward
batasa	bataasaa	sugar candy
begar	begaar	forced, or corvée labor
beta	beṭaa	male child
beti	beṭii	female child
bhaibandh	bhaibandh	brotherhood
bhakti	bhakti	love, faith
Bhangi	bhangii	sweeper
bhikku	bhikku	Buddhist monk
bhut	bhuut	ghost
bidi	biiḍii	Indian cigarette
bigha	biighaa	five-eighths of an acre of land
bodhisatva	bodhisatva	an enlightened Buddhist
bohare	boharaa; byoharaa	money lender

Transliteration	*Hindi*	*Gloss*
bopari	bopaarii; byopaarii	retailer, businessman
Brahman	braahman	Brahman, name of Hindu caste
caca	caacaa	paternal uncle
caci	caacii	paternal aunt
Camar	camaar	name of a caste group, leather worker
caurasi	cauraasii	eighty-four
charma kara	carma kaara	leather worker
chaudhury	caudharii	headman
crore	karor	ten million
culha	cuulhaa	oven
dada	daadaa	paternal grandfather
dadi	daadii	paternal grandmother
dal	dal	group, clique
daliawala	daliyaavaalaa	one who carries a basket on his head
darshana	darshana	view, sight, seeing
dehari	deharii	doorsill, threshold
deshi	deshii	pertaining to the country
dhamma	dhamma	religion, duty
Dhobi	dhobii	washerman, name of a caste
diksha	diikshaa	initiation
dushera	dasharaa	Hindu feast day
gaekwar	gaikwar	a ruler of Maharashtra
gauna	gaunaa	beginning of cohabitation
ghar	ghar	house
gharwala	gharvaalaa	house dweller
ghunghat	ghuunghat	face veiling
gotra	gotra	lineage
guru	guru	teacher, guide
haldi	haldii	tumeric
Harijan	harijan	Untouchable
Hindu	hinduu	Hindu
Holi	holii	a Hindu festival
hookah	hukkaa	water pipe
insan	insaan	humanity

Transliteration	Hindi	Gloss
jagriti	jaagṛti	awakening, enlightenment
jajman	jajmaan	patron
jajmani	jajmaanii	of patron-client system
jat	jaṭ	camel driver
Jatav	Jaaṭav	name of a caste, leather worker
jati	jaati	sub-caste, caste
jyonar	jyonaar	feast, banquet
kaccha	kaccaa	raw, crude, imperfect
karigar	kaariigaar	craftsman, artisan
karobar	kaarobaar	factory owner
katauti	katautii	discount, small subtracted amount
Khatik	khaṭik	name of a caste
kirayedar	kiraayedaar	renter, tenant
Kshatriya	kshatriya	Kshatriya, man of warrior caste
Kumhar	kumhaar	potter, man of potter caste
Kuril	kuriil	name of a caste, leather worker
lakh	laakh	one hundred thousand
Lodhi	lodii	name of a caste
log	log	people
Lohar	lohaar	name of a caste, ironsmith
mahan	mahaan	great, eminent
Mahar	mahaar	name of a caste
mahatma	mahaatmaa	a great sage
maidan	maidaan	field, open space
makan	makaan	house
Mallah	mallaah	name of a caste, boatman
malwala	maalvaalaa	one who sells supplies (lit., things-person)
mandal	maṇḍal	society, circle
mandi	maṇḍii	market, bazaar, place
mara	maaṛaa	a canopy, feast of marriage
marg	maarg	boulevard, road
mazdoor	majduur	workman, laborer

Transliteration	*Hindi*	*Gloss*
mela	melaa	a fair
mistri	mistrii	master craftsman
mahalla	mahallaa, muhallaa	ward, neighborhood of a city
mora	moraa	male child
mori	morii	female child
Nai	naaii	barber
namewala	naamevaalaa	one who sells on order (lit., book-person)
nazul	najuul	government land
nyota	nyotaa, nyautaa	invitation
pakka	pakkaa	good, well-made
pan	paan	betel leaf
panch	panc	member of a panchayat
panchayat	pancaayat	caste or village council
Parasuram	parashuraam	name of a mythical Brahman
parhe-likhe	parhe-likhe	educated, literate
peshgar	peshgaar	one who presents the docket in a court
phul	phuul	bones, remains of cremated person
piao	piyaav	drinking stand
pir	piir	saint
punji pati	puunjii pati	capitalist, rich man
purjan	parjan	dependent, servant
raj	raaj	rule, government
raksha-bandhan	rakshaa-bandhan	Hindu festival for brothers and sisters
ras	raas	a play
Reghar	reghar	name of a caste, leather worker
roti	rotii	flat break
rozi	rojii	work, employment
rupee	rupyaa	a rupee, Indian dollar
sainik	sainik	guardian, soldier
sakhaa	saakhaa	spigotted water can
samadhi	samaadhi	contemplation, insight
samuday	samudaay	collection, set, community

Transliteration	Hindi	Gloss
sanatani	sanaatanii	orthodox, traditional
sandesh	sandesh	message, news
sanskar	sanskaar	counsel, sacrament, rite
sarpanch	sarpanc	head of a panchayat
satsang	satsangh	group of pious men
satta	sattaa	authority, power
satyagraha	satyaagraha	civil disobedience
seth	seth	wealthy person
shadi	shaadii	marriage ceremony
shahid	shahiid	martyr
Shudra	shuudra	a man of the fourth and lowest Hindu class
svaran	svaraṇ = savarṇa	upper caste, of same caste
swami	svaamii	teacher
tahsildar	tahsiildaar	revenue collector
Thakur	ṭhaakur	name of a caste, warrior
thok	thok	part of a village or city ward, amount of something
thokwala	thokvaalaa	member of a thok
topi	ṭopii	hat
trepan	trepan	fifty-three
Vaisya	vaishya	third of the Hindu classes, the merchant class
Valmiki	vaalmiiki	name of a caste, sweeper
vansh	vaṇsh	race, clan
varna	varṇa	Hindu class (social)
varanashram	varṇaashram	one's caste, theory of caste society
vidhan	vidhaan	law, rule
vihara	vihaara	Buddhist temple or monastery
vyapari	vyaapaarii	businessman

II. PHRASES

Transliteration	*Hindi*	*Gloss*
Adi Hindu Andolan	aadi hinduu aandolan	original Indian movement
Amar Ujala	amar ujaalaa	eternal light
Ambedkar Jayanti	Ambedkar jayantii	Ambedkar's birthday
Arya Samaj	aarya samaaj	Aryan Society
Ashok Vijaydashmi	ashok vijaydashmii	Ashoka's Victory Day
Baba Sahab	baabaa saahab	elderly lord, respected sire
bare admi	bare aadmii	big men
bare petwala	bare peṭwaalaa	rich man (lit., big stomached one)
Bharatiya Bodh Mahasabha	bhaaratiiya bodh mahaasabhaa	Indian Buddhist Association
bhaumver phere	bhaamvar phere	circling the sacred fire at marriage
Bodh Dharam	bodh dharam	Buddhist Religion
Bodh Jayanti	bodh jayantii	Buddha's Birthday
Bodh Sammelan	bodh sammelan	Buddhist Conference
braj bhasha	braj bhaashaa	dialect of Hindi
chari bardar	charii bardaar	mace bearer
Dalit Varg sangh	dalit varg saṅgh	Depressed Classes League
Gyan Samudra	gyaan samudra	Ocean of Knowledge
Jai Bhim	jay bhiim	Hail Bhim
Jatav Jivan	jaaṭav jiivan	Jatav Life
Jatav Pracharak Mandal	jaaṭav pracaarak maṇḍal	Jatav Propaganda Circle
Jatav Vir Mahasabha	jaaṭav viir mahaasabhaa	Jatav Men's Association
Jatav Yuvak Mandal	jaaṭav yuvak maṇḍal	Jatav Young Men's Circle
Jat-Pat Todak Mandal	jaat-paant toṛak maṇḍal	Casteism Eradication Circle

Transliteration	Hindi	Gloss
Kela Devi	kelaa devii	goddess Kela
krantikari daur	kraantikaarii dauṛ	revolutionary campaign
Krishna Janamastmi	Kṛshṇa janamashṭmii	Krishna's birthday
Lok Sabha	lok sabha	Peoples' Council
mook nayak	muuk naayak	leader of the dumb
Nau Jagriti	nau jaagṛti	New Awakening, Enlightenment
naye paise	naye paise	new cents
neta log	netaa log	leader people, leaders, politicians
nili topi	niilii ṭopii	blue cap
Yaduvansh ka Itihas	yaaduvaṇsh kaa itihaas	History of Yadav Race
Vidhan Sabha	vidhaan sabhaa	Provincial Legislature
Zamin ke Tare	jamiin ke taare	Stars of the Land

APPENDIX

TABLE A–I. *Monthly Wages of Ten Workers in an Organized Factory (April 1963 to March 1964)*

April	May	June	July	August	September	October
00.00	27.75	85.50	120.25	103.00	78.50	109.50
138.75	154.50	107.25	161.54	159.25	150.00	187.25
46.22	46.81	74.88	45.72	74.00	45.50	78.73
152.99	165.88	154.19	208.03	217.41	148.92	239.37
135.88	30.72	167.57	155.32	138.23	153.33	140.26
88.61	74.47	27.36	167.16	114.64	117.63	172.19
271.77	241.64	215.17	230.96	314.10	277.97	266.95
182.09	213.15	212.28	239.14	129.64	218.39	394.69
170.25	203.51	236.00	273.24	135.24	163.25	315.00
243.20	341.73	501.00	359.29	321.00	228.75	343.50
1429.76	1500.16	1781.20	1960.65	1707.51	1582.24	2247.44
142.98	150.02	178.12	196.07	170.75	158.22	224.74 monthly average

November	December	January	February	March	Total
93.00	93.00	162.00	117.00	117.00	1106.50
123.25	00.00	00.00	276.00	220.25	1678.04
67.75	48.00	24.25	30.00	57.00	638.86
295.18	191.16	281.99	259.84	208.19	2523.15
109.45	197.38	194.12	217.60	140.28	1781.14
227.16	231.10	310.05	120.76	00.00	1651.13
376.78	245.07	298.28	262.51	230.60	3231.81
425.08	279.90	244.72	202.20	420.29	3161.56
195.75	228.00	353.00	320.25	225.00	2818.49
279.00	174.00	231.00	234.00	93.00	3439.37
2192.40	1687.61	2189.41	2040.16	1711.61	22030.15
219.24	168.76	218.94	204.01	171.16	monthly average

Average Annual Wage in Rupees 2203.02
Average Monthly Wage in Rupees 183.75

TABLE A–II. *Monthly Wages of Ten Workers in an Unorganized Factory (August 1963 to July 1964)*

August	September	October	November	December	January	
146.00	86.00	130.00	32.00	20.00	80.00	
18.00	36.50	27.00	26.00	21.00	12.00	
14.00	35.00	62.50	53.25	132.00	177.00	
00.00	28.75	52.50	34.00	72.00	98.00	
13.00	95.00	95.00	42.00	128.00	123.00	
00.00	96.00	71.00	45.00	61.00	113.00	
5.00	84.00	123.00	107.50	78.00	61.25	
5.00	68.00	58.50	4.50	30.75	11.00	
7.00	3.00	12.00	32.00	54.00	53.00	
43.00	102.35	136.67	59.50	39.05	92.00	
251.00	634.60	768.17	435.75	635.80	820.25	
25.10	63.46	76.82	43.58	63.58	82.03	monthly average

February	March	April	May	June	July	Total
30.00	149.00	137.00	282.00	120.00	00.00	1302.00
57.00	49.00	41.00	10.00	10.00	5.00	312.50
87.81	117.00	156.50	78.50	17.00	49.00	979.56
51.50	70.25	42.00	70.25	00.00	00.00	519.25
86.00	29.00	10.00	66.00	20.00	9.00	716.00
78.00	67.00	13.00	142.00	00.00	5.00	691.00
49.75	82.75	61.25	15.00	12.00	00.00	679.50
25.00	19.00	35.00	26.00	8.00	00.00	290.75
50.00	45.00	30.00	80.00	15.00	20.00	401.00
136.25	149.75	12.00	00.00	10.15	5.00	785.72
651.31	777.75	537.75	769.75	212.15	183.00	6677.28
65.13	77.78	53.78	76.98	21.22	18.30	monthly average

Average Annual Wage in Rupees   667.73
Average Monthly Wage in Rupees   55.65

TABLE A–III. *Population of Bhim Nagar*

Group	Male	Female	Total
Ten Thoks*			
A (1 and 2)	300	253	553
B	61	68	129
C	139	87	226
D	94	111	205
E	35	27	62
F	37	24	61
G	73	55	128
H	110	80	190
I	55	46	101
Total Ten Thoks	904	751	1,655
Other Jatavs			
Outsiders	61	45	106
Neighbors	62	47	109
Total Other Jatavs	123	92	215
Total all Jatavs	1,027	843	1,870
Other Castes			
Sweeper	16	17	33
Muslim	34	31	65
Merchant	2	1	3
Koli	16	14	30
Washerman	3	3	6
Sindhi	30	21	51
Total Other Castes	101	87	188
Grand Total	1,128	930	2,058

* A *thok* is a sub-unit of a *neighborhood*. There are ten such sub-units in Bhim Nagar. *Thok* A, while it generally acts as a single unit, is generally referred to as two thoks. This is because some sister's sons (*bhanjas*) came to settle there, and as a matter of deference and courtesy they were and are nominally at least considered to form a separate *thok*.

TABLE A–IV. *Family Types**

	A	B	C	D	E	F	G	H	I	J	K	L	Total
Jatavs													
Ten Thoks	127	48	2	16	34	7	8	4	8	5	1ᵃ	10	270
Neighbors	1	5	0	2	1	1	0	1	1	0	0	3	15
Outsiders	15	4	0	0	0	0	0	0	2	0	0	0	21
Total Jatav	143	57	2	18	35	8	8	5	11	5	1	13	306
Type Total	220				72						1	13	306
Other Castes													
Sweepers	5	1	0	0	0	0	0	0	0	0	0	0	6
Koli	6	0	0	0	0	0	0	0	0	0	0	1	7
Sindhiᵇ	2	2	0	0	1	0	1	0	1	0	0	0	7
Merchant	0	1	0	0	0	0	0	0	0	0	0	0	1
Muslims	7	2	0	1	1	0	0	0	0	0	0	0	11
Washermen	0	0	0	2	0	0	0	0	0	0	1ᶜ	0	3
Total Other Castes	20	6	0	3	2	0	1	0	1	0	1	1	35
Type Total	29				76						2	14	341

ᵃ A widow who would not cooperate; probably a single member.
ᵇ All the Sindhis live in one large *pakka* house.
ᶜ This is a polygamous household with a nephew as a child.

A Nuclear	*H* Supplementary Lineal Joint
B Supplementary Nuclear	*I* Supplementary Collateral Joint
C Nuclear Polygamous	*J* Supplementary Lineal-Collateral
D Sub-Nuclear	Joint
E Lineal Joint	*K* Other
F Collateral Joint	*L* Single-Person Household
G Lineal-Collateral Joint	

* All of these types have been taken from Kolenda (1968) except for C. Nuclear Polygamous. We have not followed her in taking the residence as a joint family. Our minimal criterion was shared expenses. Kolenda defines her categories thus:

  "1. *Nuclear family:* a couple with or without unmarried children.
  2. *Supplemented nuclear family:* a nuclear family plus one or more unmarried, separated, or widowed relatives of the parents, other than their unmarried children.
  3. *Subnuclear family:* a fragment of a former nuclear family. Typical examples are the widow with unmarried children, or the widower with unmarried children, or siblings—whether unmarried, or widowed, separated, or divorced—living together.
  4. *Single-person household.*
  . . . .
  6. *Collateral joint family:* two or more married couples between whom there is a

TABLE A–V. *Education by Sex in Bhim Nagar (Jatavs Only)*

Education Level (Years)	Male	Female
1	48	20
2	39	6
3	30	9
4	45	8
5	52	8
6	21	4
7	12	1
8	13	1
9	10	1
10	6	0
11	4	0
12	4	0
13 B.A. (Prev) or B.Sc.	2	0
14 B.A. (Final)	1	0
B.A. (Graduate)	1	0
M.A. (Graduate)	1	0
Diploma Students in Indian Institute of Technology at Agra	3	0

NOTES TO TABLE A–IV *(continued)*

sibling bond—usually a brother-sister relationship—plus unmarried children.

7. *Supplemented collateral joint family:* a collateral joint family plus unmarried, divorced, or widowed relatives. Typically, such supplemental relatives are the widowed mother of the married brothers, or the widower father, or an unmarried sibling.

8. *Lineal joint family:* two couples between whom there is a lineal link, usually between parents and married son, sometimes between parents and married daughter.

9. *Supplemented lineal joint family:* a lineal joint family plus unmarried, divorced, or widowed relatives who do not belong to either of the lineally linked nuclear families; for example, the father's widower brother or the son's wife's unmarried brother.

10. *Lineal-collateral joint family:* three or more couples linked lineally and collaterally. Typically, parents and their two or more married sons, plus the unmarried children of the three or more couples.

11. *Supplemented lineal-collateral joint family:* a lineal-collateral joint family plus unmarried, widowed, separated relatives who belong to none of the nuclear families lineally and collaterally linked; for example, the father's widowed sister or brother, or an unmarried nephew of the father." (Kolenda 1968: 346–47).

BIBLIOGRAPHY

Works in English

All India Jatav Youth League. *Memorial to the Most Honorable Marquess of Zetland.* Agra, 1938.

Almond, Gabriel A. "Introduction: A Functional Approach to Comparative Politics" in *The Politics of the Developing Areas,* edited by Gabriel A. Almond and James S. Coleman. Princeton: Princeton University Press, 1960, pp. 3–64.

Ambedkar, B. R. *The Buddha and His Dhamma.* Bombay: Siddharth College Publication, 1957.

———. *The Untouchables: Who Were They and Why They Became Untouchables.* New Delhi: Amrit Book Company, 1948.

———. *What Congress and Gandhi Have Done to the Untouchables.* Bombay: Thacker and Co., Ltd., 1946.

———. *Annihilation of Caste.* Bombay: The Bharat Bhusan P. Press, 1945.

Asvaghosa. *The Vajrasuci: Sanskrit Text,* edited and translated by Sujitkumar Mukhopadhyaya. Santiniketan: The Sino-Indian Cultural Society, 1950.

Bailey, F. G. "Closed Social Stratification in India," *Archives Europeennes de Sociologie,* IV (1) (1963), pp. 107–24.

———. *Tribe, Caste and Nation.* Manchester: Manchester University Press, 1960.

———. *Caste and the Economic Frontier.* Manchester: Manchester University Press, 1957.

Barber, Bernard. *Social Stratification.* New York: Harcourt, Brace and Co., 1957.

———. "Social Mobility in Hindu India" in *Social Mobility in the Caste System of India,* edited by James Silverberg (*Comparative Studies in Society and History,* Supplement III). The Hague: Mouton, 1968.

Barth, Fredrik. "The System of Social Stratification in Swat, North Pakistan" in *Aspects of Caste in South India, Ceylon and Northwest Pakistan,* edited by E. R. Leach (Cambridge Papers in Social Anthropology, No. 2). Cambridge: Cambridge University Press, 1960, pp. 113–46.

Berreman, Gerald D. "Structure and Function of Caste Systems" in

Japan's Invisible Race, edited by George DeVos and Hiroshi Wagat-
suma. Berkeley: University of California Press, 1966, pp. 277–307.
———. "The Study of Caste Ranking in India," *Southwestern Journal
of Anthropology,* XXI (2) (1965), pp. 115–29.
———. "Aleut Reference Group Alienation, Mobility, and Accultura-
tion," *American Anthropologist,* LXVI (2) (1964), pp. 231–50.
Beteille, Andre. *Caste, Class, and Power.* Berkeley and Los Angeles:
University of California Press, 1965a.
———. "The Future of the Backward Classes: the Competing Demands
of Status and Power," *Perspectives, Supplement to the Indian Jour-
nal of Public Administration,* XI (1) (1965b), pp. 1–39.
Bhanu, Dharma. *History and Administration of the North-Western
Province (Subsequently called the Agra Province), 1800–1858.* Agra:
S. L. Agarwala and Co., Pvt., Ltd., 1957.
Bharatiya Juta Grih Utpadak Sangh. *Problems of Footwear Industry:
The Memorandum Submitted to the Planning Commission.* Agra:
Bansal Press, n.d.
Bhatt, G. S. "Urban Impact and the Changing Status of the Chamars
of Dehra Dun." Paper presented at Indian Sociological Conference.
Saugar [India], 1960. (Typewritten.)
Bopegamage, A. *Delhi: A Study in Urban Sociology.* (University of
Bombay Publications, Sociology Series No. 7). Bombay: University
of Bombay, 1957.
Brass, Paul R. *Factional Politics in an Indian State: The Congress
Party in Uttar Pradesh.* Berkeley and Los Angeles: The University
of California Press, 1965.
Briggs, George W. *The Chamars.* Calcutta: Association Press, 1920.
Burn, R. "Kabir, Kabirpanthis" in *Encyclopedia of Religion and
Ethics,* edited by James Hastings. New York: Charles Scribner's Sons,
Vol. VII (1915), pp. 632–34.
Chapple, Eliot D. and Carleton S. Coon. *Principles of Anthropology.*
New York: Henry Holt and Company, 1942.
Cohn, Bernard S. "Chamar Family in a North Indian Village: A Struc-
tural Contingent," *Economic Weekly* [of Bombay], XIII (27, 28, 29)
(1961), pp. 1051–55.
———. "Some Notes on Law and Change in North India," *Economic
Development and Cultural Change,* VIII (1) (1959), pp. 79–93.
———. "The Changing Status of a Depressed Caste" in *Village India,*
edited by McKim Marriott. Chicago: University of Chicago Press,
1955, pp. 53–77.
———. "The Camars of Senapur: A Study of the Changing Status of a

Depressed Caste." Unpublished Ph.D. dissertation, Cornell University, 1954.

Cornell, John B. "From Caste Patron to Entrepreneur and Political Ideologue: Transformation of Nineteenth and Twentieth Century Outcaste Leadership Elites." Paper presented at Conference on the Nineteenth Century Elites. Tucson: 1963. (Mimeographed.)

Crooke, W. *The Tribes and Castes of the North-Western Provinces and Oudh.* Calcutta: Office of the Superintendent of Government Printing, India, Vol. II, 1896.

Dowson, John. *A Classical Dictionary of Hindu Mythology and Religion, Geography, History, and Literature.* London: Routledge and Kegan Paul, Ltd., 1961.

Dumont, Louis and D. Pocock. "Village Studies," *Contributions to Indian Sociology,* Vol. I (1957), pp. 23–42.

Dushkin, Lelah. "Special Treatment Policy," *Economic Weekly,* XIII (43) (1961a), pp. 1665–68.

———. "Special Treatment Provisions," *Economic Weekly,* XIII (44 and 45) (1961b), pp. 1695–1705.

———. "Future of Special Treatment," *Economic Weekly,* XIII (46) (1961c), pp. 1729–38.

———. "The Policy of the Indian National Congress Toward the Depressed Classes: An Historical Study." Unpublished Master's thesis, Department of South Asia Studies, University of Pennsylvania, 1957.

Economic Weekly [of Bombay].

Essien-Udom, E. U. *Black Nationalism.* New York: Dell Publishing Co., 1964. (First published in 1962: Chicago: University of Chicago Press.)

Festinger, Leon. *A Theory of Cognitive Dissonance.* Evanston: Row, Peterson, 1957.

Frazer, Thomas Mott. "Directed Change in India." Unpublished Ph.D. dissertation, Columbia University, 1963.

Friedl, Ernestine. "Lagging Emulation in Post-Peasant Society," *American Anthropologist,* LXVI (3, Pt. 1) (1964), pp. 569–86.

Fuchs, Stephen. *Rebellious Prophets: a Study of Messianic Movements in Indian Religions.* Bombay: Asia Publishing House, 1965.

Galanter, Marc. "Law and Caste in Modern India," *Asian Survey,* III (11) (1963), pp. 544–59.

———. "Equality and 'Protective Discrimination' in India," *Rutgers Law Review,* XVI (1) (1961), pp. 42–74.

Geertz, Clifford. "The Integrative Revolution" in *Old Societies and*

New States, edited by Clifford Geertz. Glencoe: The Free Press, 1963, pp. 105–57.

Goffman, Erving. *The Presentation of Self in Everyday Life.* Garden City: Doubleday and Co., Inc. (Anchor Books), 1959.

Goodenough, Ward H. "Residence Rules," *Southwestern Journal of Anthropology,* XII (1) (1956), pp. 22–37.

Gough, E. Kathleen. "The Social Structure of a Tanjore Village" in *Village India,* edited by McKim Marriott. Chicago: University of Chicago Press, 1955, pp. 36–52.

Gould, Harold A. "The Adaptive Functions of Caste in Contemporary Indian Society," *Asian Survey,* III (9) (1963), pp. 427–38.

———. "Sanskritization and Westernization," *Economic Weekly,* XIII (25) (1961), pp. 945–50.

Goyal, Prem Prakash. "An Inquiry into the Impact of Urbanism on the Magico-Religious Beliefs and Practices of Raidas Chamars of Dehra Dun." Unpublished Master's thesis, Department of Anthropology, D. A. V. College (Dehra Dun, India), 1961.

Grierson, George A. "Ramanandis, Ramawats" in *Encyclopedia of Religion and Ethics,* edited by James Hastings. New York: Charles Scribner's Sons, Vol. X (1919), pp. 569–72.

Griswold, H. D. "Arya Samaj" in *Encyclopedia of Religion and Ethics,* edited by James Hastings. New York: Charles Scribner's Sons, Vol. II (1913), pp. 57–62.

Hamilton, Walter. *East Indian Gazeteer.* London: Wm. H. Allen and Co., Vol. I, 1828.

Hoffer, Eric. *The True Believer.* New York: Harper and Row (Perennial Library), 1966.

India. Census Commission. *Census of India—1961. Uttar Pradesh General Population Tables.* Vol. XV (Pt. IIA). Delhi: Manager of Publications, 1964.

———. *Census of India—1961. Religion.* Paper No. 1 of 1963. Delhi: Manager of Publications, 1963.

———. *Census of India—1951. Uttar Pradesh.* Vol. II (Pt. 1A). Allahabad: Superintendent, Printing and Stationery, Uttar Pradesh, 1953.

———. *Census of India—1911. United Provinces of Agra and Oudh.* Vol. XV (Pt. II). Allahabad: Superintendent, Government Press, 1912.

———. *Census of India—1901. North-Western Provinces and Oudh.* Vol. XVIA (Pt. II). Allahabad: Superintendent, Government Press, 1902.

————. *Census of the North-Western Provinces and Oudh—1881.* Supplement to the Report on the Census. Vol. IX (Pt. 2). Allahabad: North-Western Provinces and Oudh Government Press, 1882.

India. Commissioner for Scheduled Castes and Scheduled Tribes. *Report of the Commissioner for Scheduled Castes and Scheduled Tribes for the Year 1960–61.* (Pt. I). Prepared by L. M. Shrikant. Delhi: Manager of Publications, 1962.

India. Development Commissioner, Small Scale Industries. *Leather Footwear (Northern Region)* (Small Scale Industry Analysis and Planning Report No. 4.). New Delhi: Ministry of Commerce and Industry, 1956.

Indian Political Science Conference. *Indian Political Science Conference: Silver Jubilee, December, 1963, Agra.* Agra: Agra University Press, 1963.

Isaacs, Harold. *India's Ex-Untouchables.* New York: John Day and Co., 1965.

Jatav-Vir (Educational) Institute. *Report of the Jatav-Vir (Educational) Institute, Agra.* Agra: Board of Management of Jatav-Vir Institute, 1939.

Keer, Dhananjay. *Dr. Ambedkar, Life and Mission.* Bombay: Popular Prakashan, 1962.

Kolenda, Pauline M. "Region, Caste, and Family Structure: A Comparative Study of the Indian 'Joint' Family" in *Structure and Change in Indian Society,* edited by Milton Singer and Bernard Cohn (Viking Fund Publications in Anthropology, No. 47). Chicago: Aldine Press, 1968, pp. 339–96.

————. "Toward a Model of the Hindu Jajmani System," *Human Organization,* XX (1) (1963), pp. 11–31.

Lanternari, Vittorio. *The Religions of the Oppressed.* New York: Mentor Books (M. T. 608), 1963.

Latif, Syad Muhammad. *Agra: Historical and Descriptive.* Calcutta: Calcutta Central Press Co., Ltd., 1896.

Leach, E. R. "Introduction: What Should We Mean by Caste?" in *Aspects of Caste in South India, Ceylon and North-West Pakistan,* edited by E. R. Leach (Cambridge Papers in Social Anthropology, No. 2). Cambridge: Cambridge Univeristy Press, 1960, pp. 1–10.

Lillingston, Frank. "Chamars" in *Encyclopedia of Religion and Ethics,* edited by James Hastings. New York: Charles Scribner's Sons, Vol. III, 1913, pp. 351–55.

Lynch, Owen M. "An Untouchable Culture Hero: Myth and Cha-

risma" in *Untouchables in Contemporary India,* edited by J. Michael
Mahar. Tucson: The University of Arizona Press (in press).
——. "The Politics of Untouchability: A Case from Agra India" in
Structure and Change in Indian Society, edited by Milton Singer
and Bernard Cohn (Viking Fund Publications in Anthropology, No.
47). Chicago: Aldine Press, 1968, pp. 209–40.
——. "Rural Cities in India: Continuities and Discontinuities" in
India and Ceylon: Unity and Diversity, edited by Philip Mason (The
Institute of Race Relations, London). London: Oxford University
Press, 1967, pp. 142–58.
——. "Some Aspects of Rural-Urban Continuum in India" in
Anthropology on the March, edited by Bala Ratnam. Madras: The
Book Center, 1963, pp. 178–205.
Mahar, Pauline M. "Changing Religious Practices of an Untouchable
Caste," *Economic Development and Cultural Change,* VII (3) (1960),
pp. 279–87.
——. "Changing Caste Ideology in a North Indian Village," *Journal
of Social Issues,* XIV (4) (1958), pp. 51–65.
Manager, Government Pilot Project. *Footwear Industry at Agra.* n.d.
Mandelbaum, David G. "Transcendental and Pragmatic Aspects of
Religion," *American Anthropologist,* LXVIII (5) (1966), pp. 1174–
91.
Marriott, McKim. "Changing Channels of Cultural Transmission in
Indian Civilization" in *Intermediate Societies, Social Mobility, and
Communication,* edited by Verne F. Ray (Proceedings of the 1959
Annual Spring Meeting of the American Ethnological Society).
Seattle: American Ethnological Society, 1959, pp. 66–74.
Marshall, T. H. *Class, Citizenship and Social Development.* Garden
City: Doubleday and Co., Inc. (Anchor Books), 1965.
Merton, Robert K. *Social Theory and Social Structure.* Glencoe: The
Free Press, 1957.
Miller, Beatrice D. "Revitalization Movements: Theory and Practice,"
paper to be published in Verrier Elwin Memorial Volume, n.d.
(Dittographed.)
Morris-Jones, W. H. *The Government and Politics of India.* London:
Hutchinson University Library, 1964.
Nadel, S. F. *The Theory of Social Structure.* London: Cohen and
West, Ltd., 1957. New York: The Free Press (A Division of The
MacMillan Company).
——. *The Foundations of Social Anthropology.* London: Cohen and
West, Ltd., 1951.

Nesfield, J. C. *Brief Review of the Caste System of the North-Western Provinces and Oudh*. Allahabad: North-Western Provinces and Oudh Government Press, 1885.

Nevill, H. R. *Agra: A Gazetteer*. (Agra and Oudh, United Provinces of District Gazetteers, Vol. 8). Allahabad: Superintendent, Government Press, 1921.

Nigh, Wilbur J. Personal communication, 1965.

Orans, Martin. *The Santal: a Tribe in Search of a Great Tradition*. Detroit: Wayne State University Press, 1965.

Pocock, David F. "Sociologies Urban and Rural," *Contributions to Indian Sociology*, Vol. IV (1960), pp. 63–81.

Prasad, Narmadeshwar. *The Myth of the Caste System*. Patna: Samjna Prakashan, 1957.

Rai, Lajpat. *The Arya Samaj*. London: Longmans, Green and Co., 1915.

Republican Party of India. *Charter of Demands*. New Delhi: n.p., 1965.

———. *Election Manifesto*. New Delhi: Ganga Printing Press, n.d.

Rosenthal, Donald B. "Factions and Alliances in Indian City Politics," *Midwest Journal of Political Science*, X (3) (1966), pp. 320–49.

———. *City Politics in India*. (Unpublished manuscript.) n.d.

Rowe, William L. "Social and Economic Mobility in a Low Caste North Indian Community." Unpublished Ph.D. dissertation, Cornell University, 1960.

Roy, Ramashray. "Congress Defeat in Farrukhabad," *Economic Weekly*, XVII (22) (1965), pp. 893–902.

Rudolph, Lloyd I. and Susan H. "The Political Role of India's Caste Associations," *Pacific Affairs*, XXXIII (1) (1960), pp. 5–22.

Sahlins, Marshall D. and Elman R. Service. *Evolution and Culture*. Ann Arbor: The University of Michigan Press, 1960.

Sharma, Kailash Baboo. "Marketing of Leather Goods at Agra." Unpublished Master's thesis, Department of Commerce, Balwant Rajput College (Agra, India), 1958.

Singer, Milton. "The Indian Joint Family in Modern Industry" in *Structure and Change in Indian Society*, edited by Milton Singer and Bernard Cohn (Viking Fund Publications in Anthropology, No. 47). Chicago: Aldine Press, 1968, pp. 423–52.

Sisodia, Ram Jhalak. "Cottage Shoe Workers in Agra." Unpublished Master's thesis, Department of Commerce, Balwant Rajput College (Agra, India), 1960.

Smelser, Neil J. "Toward a Theory of Modernization" in *Social*

Change: Sources, Patterns, and Consequences, edited by Amitai and Eva Etzioni. New York: Basic Books, Inc., 1964, pp. 258–74.

Smith, Donald. *India as a Secular State.* Princeton: Princeton University Press, 1963.

Srinivas, M. N. *Social Change in Modern India.* Berkeley: University of California Press, 1966.

———. *Religion and Society among the Coorgs of South India.* Bombay: Asia Publishing House, 1965.

———. *Caste in Modern India and Other Essays.* Bombay: Asia Publishing House, 1962.

———. "The Dominant Caste in Rampura," *American Anthropologist,* LXI (1) (1959), pp. 1–16.

———. "Caste in Modern India," *Journal of Asian Studies,* XVI (4) (1957), pp. 529–48.

———. "A Note on Sanskritization and Westernization," *Far Eastern Quarterly,* XV (4) (1956), pp. 481–96.

Srivastava, S. K. "The Process of Desanskritization in Village India" in *Anthropology on the March,* edited by Bala Ratnam. Madras: The Book Center, 1963, pp. 263–67.

Tangri, Shanti. "Urbanization, Political Stability and Economic Growth" in *India's Urban Future,* edited by Roy Turner. Bombay: Oxford University Press, 1962.

Tiwari, A. R. "Urban Regions of Agra," *Agra University Journal of Research,* VI (1) (1958), pp. 101–14.

Turner, Ralph H. "Role-taking, Role Standpoint, and Reference-group Behavior," *American Journal of Sociology,* LXI (4) (1956), pp. 316–28.

Upadhyay, Kali Charan. "Condition of Labour in Leather Industry at Agra." Unpublished Master's thesis, Department of Economics, Balwant Rajput College (Agra, India), 1951.

Wallace, Anthony F. C. *Culture and Personality.* New York: Random House, 1961.

Waterfield, William. *The Lay of Alha.* London: Oxford University Press, 1923.

Weber, Max. *The Sociology of Religion.* Translated by Ephraim Fischoff. Boston: Beacon Press, 1963.

———. *From Max Weber: Essays in Sociology.* Translated and edited by H. H. Gerth and C. Wright Mills. New York: Oxford University Press, 1958.

Weiner, Myron. *The Politics of Scarcity.* First Indian edition. Bombay: Asia Publishing House, 1963.

Wolf, Eric R. "Aspects of Group Relations in a Complex Society: Mexico," *American Anthropologist*, LVIII (6) (1956), pp. 1065–78.

Zelliot, Eleanor. "Buddhism and Politics in Maharashtra" in *Religion and Politics in South Asia,* edited by Donald E. Smith. Princeton: Princeton University Press, 1966, pp. 199–212.

————. Personal communication, 1965.

Works in Hindi

Amar Ujaalaa. [Daily newspaper, Agra, India.]

Dalit Varg Sangh. *Jaaṭav Bhaaiyõ Ko Patra.* Aagraa: Janata Press, 1962.

Jigyaasu, Chandrikaa Prasaad. *Shrii 108 Svaamii Achhuutaanandjii Harihar.* Lukhnauu: Hinduu Samaaj Sudhaar Kaaryaalay, 1960.

————. *Bhaaratiiya Ripablikan Paartii hii Kyõ Aavashyak hai.* n.p.: Bahujan Kalyaaṇ Prakaashan, n.d.

Nau Jaagriti. [Occasional newspaper, Agra, India.]

Sagar, Pandit Sundarlaal. *Jaaṭav Jiivan.* Aagraa: Jaaṭav Pracaarak Mahaamaṇḍal, 1924.

Sainik. [Daily newspaper, Agra, India.]

Yaadvendu, Raamnaaraayan. *Yaaduvansh kaa Itihaas.* Aagraa: Navyug Saahitya Niketan, 1942.

INDEX

Adaptive structure. *See* Caste; Jatav
Agra City, 2, 100, 138; climate, 20; origins, 20–21; under Moghuls, 21; under British, 21; industry, 21–22; population, 22, 31
Agra Municipal Corporation, 30, 100, 102–103, 115, 118–19, 122, 125 and *n*, 204
Agra Municipal Corporators, 102, 114, 116, 119, 124. *See* also Elections
Agrawal, Dr., 117, 122, 124
Alha, 140, 188
Ambedkar, Dr. Bhimrao Ramji, 81, 86, 87, 88 and *n*, 92, 108, 119, 150, 152–53, 157, 188, 194, 208; and Republican Party, 95–96, 124; and the Constitution, 106*n*, 136–37, 139, 148; picture of, 118; charisma, 124, 127, 138; early life, 130–31; conflict with Ghandi, 132–36; as Law Minister, 136–37; and Agra Jatavs, 137–44; anti-caste religious teachings, 141–43; and Buddhist movement in Agra, 158, 163; as a prophet, 160 and *n*
Ambedkar Forward Block, 124
Analogical identification, 81
Arya Samaj, 67, 68–69, 70, 76, 84–85, 178, 180, 212
Associations, 144 and *n*, 204, 209
Asvaghosa, 92*n*

Basket men, 40, 45, 47
Bhim Nagar, neighborhood of; history, 167; description of, 167–69; population, 169; interaction of castes, 169–72; and Buddhism, 178, 198–200; as a unit, 189; education, 192–94; changes in structure, 198–200. *See* also *Panchayat; Thoks*
Big men, 35, 45, 60–62, 81–84, 101, 111; as Jatav leaders, 77–79, 183, 186, 205. *See also* Conservatives
Black Muslims, 213*n*
Brahman, 5, 6, 7, 13–14, 23, 30, 58, 70, 72, 85, 92, 95, 139, 141, 142, 178, 181,

206, 212; as a negative reference group, 10, 68, 74, 84–85, 87, 92–95
Bridge actions, 91 and *n*, 103, 147, 162, 193, 194, 195, 209, 214
British, 5, 6, 61, 78, 79*n*, 82, 84, 85, 108, 132, 133, 139, 205. *See also* Agra City
Buddhism: bridging function, 144, 165; obstacles to, 146–50; as revitalization movement, 164 and *n*; and government, 165. *See also* Ambedkar, Dr. Bhimrao Ramji; Republican Party
Buddhists: population in Agra, 30, 146; monks, 60, 142, 148, 153, 155, 156, 157; All-India Conference, 121, 152–56, 175, 189; conversion, 130, 145, 146, 147, 156–57, 162; temples, 145–46, 151, 198; Indian Buddhist Society, 146; initiation, 156. *See also* Bhim Nagar; Jatav; Marriage

Camar, 2, 29, 30, 35, 57, 75, 202; status of, 62, 63, 79, 90, 104, 200. *See also* Jatav
Caste, 70–71, 91, 94, 115, 221; adaptation of, 3, 98, 203, 204, 207–08, 209; definition of, 4*n*; dominant, 5, 11, 67*n*, 96*n*, 215 and *n*; internal relations, 8, 166, 175, 201, 215; external relations, 8, 121, 175, 183, 186, 204, 215; norms, 14, 90 and *n*, 124; *See also* Political participation
Caste mobility, 7, 71, 75, 77
Caste system: interaction and change, 3–4, 12, 14; as defined by Bailey, 10–12. *See also* Jatav; Sanskritization
Census Commissioner, 82
Chand, Bohare Khem, 78–79
Christians, 36*n*, 76, 122, 158, 181
Circle for the Elimination of Casteism, 141
Citizenship, 14, 44, 61, 91, 95, 113, 115, 139, 202, 205, 209, 213. *See also* Jatav
Civil disobedience, 87–88, 109, 121, 148
Class, 10, 35, 94, 97; definition, 12; norms, 90 and 90*n*, 124
Cognitive dissonance, 98

College students, 57, 177
Communal award, 133
Communalism, 101, 102, 109, 117, 121
Communications, 44–45, 79, 102, 115, 120–21, 192, 198
Communism, 96n, 105–106, 207
Conflict, 11, 13, 75, 80, 85, 90, 113, 158, 212; defined, 11. *See also* Dichotomization
Congress Party, 2, 10, 60, 62, 87, 93, 95, 101, 111–14, 115, 116, 123, 125, 133, 136, 137, 207; as a negative reference group, 93, 108. *See also* Scheduled Castes
Conservatives, 80, 89, 111–14, 121n, 205
Constitution of India, 7, 25, 26, 89, 90, 96, 164, 195, 206, 207. *See also* Ambedkar, Dr. Bhimrao Ramji
Contractors, 33, 40, 45, 47, 51, 67, 72
Contradiction, 11–12, 13, 14, 75, 85, 90, 127, 212, 213–14; defined, 11. *See also* Dichotomization
Controlling status, 12n, 14n, 15, 75, 80, 90, 110, 127, 147, 214, 218; defined, 14
Cooperation, 10, 11 and n, 105, 114
Corruption, 122–23, 124, 125, 192
Councils. *See Panchayat*
Country market, 40
Courts, 78, 119, 147n, 186, 194–96, 200, 211, 214
Cottage worker. *See* Jatav as shoemakers
Craftsman. *See* Jatav as shoemakers
Culture brokers, 77, 205

Depressed classes, 107n, 131–36; League, 115
Dichotomization, 16–19, 61, 63, 92, 183, 200–202, 208, 215–16; defined, 16; conflict and contradiction, 17
Differentiation, 17, 186, 215–16
Discrimination, 23, 26, 28, 39, 67, 75, 90, 91, 97, 209, 210
Dominant status, 13–15, 39, 44, 61, 62, 63, 71, 75, 80, 90, 91n, 97, 103, 110, 113, 115, 127, 212, 213–14

Education, 3, 69, 76, 79, 107, 113, 125, 156, 177–78, 192. *See also* Bhim Nagar; Protective discrimination
Elections, 88, 98, 101, 124, 132, 135; for Municipal Corporation, 100, 122; Dep-

uty Mayor, 102, 115, 117, 204, 208; Mayor of Municipal Corporation, 115; in Republican Party, 12–23
Election Manifesto. *See* Republican Party
Emulation: lagging, 215; elite, 218; solidarity conflict, 219–22
Eta, 111n
Ethnography, 2, 203; defined, 2n
Ethnology, 3, 203, 211; defined, 2n

Factors, 35–40, 42–44, 47, 59, 60, 62, 64, 105
Factory. *See* Shoe industry
Family, 36, 47, 64, 172–75; types, 172–73; attitudes in, 201. *See also*, Kinship; Marriage
Franchise, 2, 89, 139, 186, 200, 202, 204, 208

Gaekwar of Baroda, 131
Gandhi, Mahatma, 31, 81, 88 and n, 91, 132–36, 137, 138, 143, 194. *See also* Ambedkar, Dr. Bhimrao Ramji
Gandhism, 133, 136, 139
Gotras, 68, 71 and n
Guliyas, 75
Guru, 47, 48, 56, 64, 94

Harijan, 31; Welfare Office, 18, 92; Welfare Officer, 90, 91 and n, 214
Headman. *See* Jatav
Hindu, 5, 31n, 36n, 85n, 87, 101, 132, 136, 156, 221, 222; as a negative reference group, 162–63. *See also* Jatav
Hinduism, 68n, 70, 132, 133, 137, 141, 146, 165, 208; and Buddhism, 159–63
Hing Ki Mandi, 39–40

Independence in India, 1, 7, 25, 37, 70, 86, 89, 207, 222
Independents, 103, 115
Interaction, 35, 38, 63, 66, 80, 110, 114, 127, 169, 198. *See also* Caste system
Integration, 16, 17, 18, 43–44, 60, 63, 92, 120, 186, 187, 200, 201, 207, 208, 211, 215, 216, 222; mechanisms of, 187, 192, 198; and solidarity, 109–10, 201. *See also* Dichotomization
Involute system, 10, 12

Jajamini. *See* Patron-client relationship
Jan Sangh, 10, 93, 94, 103, 115, 116